New Frontiers: Modern Perspectives on Our Solar System
Part I

Professor Frank Summers

THE TEACHING COMPANY ®

PUBLISHED BY:

THE TEACHING COMPANY
4151 Lafayette Center Drive, Suite 100
Chantilly, Virginia 20151-1232
1-800-TEACH-12
Fax—703-378-3819
www.teach12.com

Copyright © The Teaching Company, 2008

Printed in the United States of America

This book is in copyright. All rights reserved.

Without limiting the rights under copyright reserved above,
no part of this publication may be reproduced, stored in
or introduced into a retrieval system, or transmitted,
in any form, or by any means
(electronic, mechanical, photocopying, recording, or otherwise),
without the prior written permission of
The Teaching Company.

ISBN 1-59803-437-5

Frank Summers, Ph.D.

Space Telescope Science Institute

Dr. Frank Summers is an astrophysicist at the Space Telescope Science Institute (STScI) in Baltimore, Maryland. He received his undergraduate degree summa cum laude in Physics from Virginia Tech and his Master's and Doctoral degrees in Astronomy from the University of California at Berkeley. Before joining the STScI, he researched cosmology and galaxy formation at Princeton University and Columbia University, specializing in supercomputer simulations, and then shifted his career emphasis to the public presentation of astronomy while at the Hayden Planetarium in New York.

His work in the Office of Public Outreach at the STScI presents the findings of the Hubble Space Telescope and developments in general astronomy to the public through news media, websites, educational programs, and exhibits at museums and planetariums. As one of only a few professional astronomers working in outreach, Dr. Summers has expertise that spans a broad range: from public astronomy presentation and scientific visualization to high-performance computing and research astrophysics. Dr. Summers also lectures widely to public audiences, students, and teachers training to teach at all educational levels.

One of Dr. Summers's particular specialties is the combination of research data and supercomputer simulations with Hollywood-level animation and rendering techniques to create accurate and aesthetic scientific visualizations. Of note, Dr. Summers directed, co-wrote, and created the 3-D visualizations for the IMAX short film *Hubble: Galaxies Across Space and Time*, which won the Large Format Cinema Association's Best Short Film Award in 2004. For the Academy Award–nominated IMAX film *Cosmic Voyage* (1996: Best Documentary, Short Subjects), he provided supercomputer data sets, assisted the scientific visualization team on a four-minute sequence, and served on the film's scientific advisory committee. He also has consulted on numerous books and television documentaries, and he can be seen on the Internet in a video podcast called *Hubble's Universe*.

Formerly, Dr. Summers was the lead astronomer in developing the exhibits for the opening of the Rose Center for Earth and Space at the American Museum of Natural History. It was during this rebirth of the museum's Hayden Planetarium that he and his fellow curators developed and presented the modern view of the solar system espoused in this course. They were criticized by some at the time, but their intellectual boldness was vindicated by the official astronomy reclassification of Pluto almost a decade later.

Table of Contents
New Frontiers:
Modern Perspectives on Our Solar System
Part I

Professor Biography		i
Course Scope		1
Lecture One	The 21st-Century Solar System	3
Lecture Two	Geometry of the Heavens	19
Lecture Three	Truth, Beauty, and Heliocentrism	35
Lecture Four	Deducing the Laws of Motion	50
Lecture Five	Planetary Predictions and Scientific Theory	65
Lecture Six	From Observational Evidence to Discoveries	79
Lecture Seven	The Space-Age Solar System	95
Lecture Eight	The Star of Our Solar System	109
Lecture Nine	Planetary Surfaces and Natural Wonders	124
Lecture Ten	Craters, Impacts, and Cataclysms	137
Lecture Eleven	Journey to the Centers of Planets	153
Lecture Twelve	Structure and Behavior of Atmospheres	171
Timeline		187
Glossary		192
Biographical Notes		205
Bibliography		211

New Frontiers:
Modern Perspectives on Our Solar System

Scope:

Since 1990, the understanding of our solar system has changed fundamentally. Recent discoveries have uncovered a whole new class of objects among our planets, as well as the existence and details of planetary systems around other stars. This knowledge grants us new perspectives on how our solar system is structured and where it stands within its larger astronomical setting.

This course explores the development of our knowledge of the solar system from the simple pondering of the motions in the sky to the latest discoveries of modern astrophysics. Our ideas about the movements of the Sun, Moon, stars, and planets have expanded as astronomers moved from human-eye examination to telescope observations to space-age exploration. Astronomy provides diverse opportunities for stunning imagery seen from ground-based observatories, space telescopes, and satellite missions to other planets. Every time we uncover new knowledge about our solar system, our viewpoint of how we see the heavens—and how we see ourselves in relation to the universe—changes.

In the first section of the course, we develop the foundations of astronomy that underlie our views of the solar system. Armed only with a keen power of observation and basic mathematics, one can develop scientific models of the motions in the heavens. For the solar system, both the geocentric (Earth-centered) and heliocentric (Sun-centered) views are justifiable hypotheses at the human-eye level of accuracy. The addition of telescopic observations both provides and demands a higher level of precision, while more advanced math allows us to deduce the laws of planetary motion. Greater understanding arises not only from painstaking hours in the dead of night at a telescope but also through careful modeling and detailed calculations. This continuing interplay of observation and interpretation has built up our modern scientific model of the solar system.

The second section of the course explores our planetary neighborhood as revealed to us through the stunning advances of the space age. Both human and robotic explorers have shown the planets, moons, asteroids, and comets of the solar system to be worlds unto themselves—strikingly different from, yet intrinsically similar to, our own. Comparative studies of such planetary characteristics as terrain, climate, weather, and internal structure show us the true extremes of nature, as well as entirely new phenomena never imagined on Earth. We will encounter subarctic cold, infernal heat, airless desolation, gigantic storms, mountains that dwarf Mount Everest, and valleys that make the Grand Canyon seem downright puny. By learning the diversity of characteristics and physical processes occurring across the solar system, we gain invaluable insight into the past and future development of our own planet.

The current revolution in our knowledge is covered in the third section of the course. New telescopes have observed thousands of icy worlds beyond Neptune and uncovered a new region of the solar system called the Kuiper Belt. Indeed, Pluto's status as a planet (as well as the very definition of a "planet") has been challenged by the discovery of Kuiper Belt objects that rival or exceed Pluto in size. Looking beyond our star, we have begun to see dusty disks collapsing to form planets around newborn stars, and for the first time in history we have detected planets orbiting other stars. Our current census of nearby stars shows many planetary systems, none of which resemble our own. Studying the properties of other planets provides insights into how planetary systems develop in general and perspectives on whether our solar system is ordinary or somehow special. In particular, we will discuss the implications for the possibility of other life in the universe. Astronomy has entered a new age of solar system exploration: one in which we have new views of both how our solar system is structured and how it compares to interstellar ideas of planetary systems.

Throughout the course, the material will be presented in a visually rich and intellectually engaging manner. The lectures also are enlivened by explanatory diagrams, illustrations, and scientific animations. The use of mathematics is nonessential to the course, though those with a basic appreciation for geometry and algebra may gain some extra insights. What *is* essential, however, is a strong curiosity and sense of wonder, as the course challenges you to walk in the shoes of those who have uncovered the secrets of the universe.

Lecture One
The 21ˢᵗ-Century Solar System

Scope:

The study of our solar system has developed over several thousand years. Currently, we are not only vastly expanding our knowledge of its contents but also changing our perspective on how those contents are organized. The solar system has had several historical changes in perspective, and this new shift has been developing over the past several decades. The old focus, on the Sun and the nine planets, has given way to a modern understanding of six categories, or families, of objects. Of particular note are the two new regions of the outer solar system, one of which precipitated the demotion of Pluto from the ranks of the planets. To fully develop an understanding of this 21ˢᵗ-century view of the solar system, this course is structured in three sections: (1) the observational and mathematical foundations of our knowledge of the solar system, (2) the detailed description of planetary worlds arising from space-age exploration, and (3) the recent revolution in understanding driven by new observations of the outer solar system and of planetary systems around other stars. As to philosophical motivation, this course both derives from and strives to engender our innate curiosity about the sky.

Outline

I. Understanding the solar system is a continuing journey of discovery.
   A. It is common to think that we already know everything important about the solar system.
      1. Science is never static, and we are always learning new ideas, both small and large.
      2. The 2006 decision to change Pluto's status illustrates both the reexamination of old knowledge and the incorporation of new knowledge.
   B. Misconceptions can get in the way of new knowledge.
      1. Many misconceptions have been created through oversimplification of ideas, selective interpretation of facts, and storytelling.

       2. It is important to carefully build a mental model that correctly interprets old facts and allows the smooth addition of novel ideas.
   C. The modern astronomical view provides not only new discoveries but also new perspectives on how to understand the content, structure, and formation of our solar system.

II. Our perspective on the solar system has changed several times in history.
   A. The Earth-centered view of the ancients gave way to the Sun-centered view of the Renaissance.
   B. As science and mathematics developed, we recognized gravity's dominant role in solar system motions.
   C. The space age enabled detailed studies and marked a change in our viewpoint, from long-range observers to in situ explorers.
   D. The 21st-century perspective is an enhanced view of the structure of the solar system as well as its universality among other planetary systems.

III. The way we teach the solar system has changed several times in my lifetime.
   A. The traditional emphasis was on the Sun and nine planets, with some mention of moons, asteroids, and comets.
       1. This approach often leads to the memorization of facts, without providing sufficient context.
       2. Many misconceptions arise from overly simplistic representations, especially the prevalent, but fundamentally flawed, depiction of the planets in equally spaced orbits and all lined up.
   B. An improved approach is called "comparative planetology."
       1. Space-age explorations provide rich details of other worlds.
       2. Comparison and contrast of these worlds to each other, and more importantly to Earth, provides a sense of how concepts apply across the solar system.
       3. The important messages are that what happens on Earth happens elsewhere and that such comparisons make the solar system understandable.

C. The 21st-century view of the solar system organizes objects into six families by their similar properties and characteristics.
  1. The Sun is a singular object, as it is the only star in the solar system.
  2. The rocky planets—Mercury, Venus, Earth, and Mars—are medium sized with small and mostly circular orbits, and few moons.
  3. The asteroid belt consists of small to tiny rocks (with some ice) that are on more elliptical and more tilted orbits, mainly in the region between Mars and Jupiter.
  4. The giant planets have deep gaseous atmospheres; rings; many moons; and large, nearly circular orbits.
  5. The Kuiper Belt is a newly observed region beyond Neptune where thousands of small to tiny icy objects orbit in sometimes rather tilted and elliptical orbits. It is the reservoir of the short-period comets.
  6. The Oort Cloud is a roughly spherical region up to a thousand times more distant than the planets that has not yet been directly observed. However, its existence is required to explain the abundance of long-period comets and their highly elongated and omnidirectional orbits.

D. Big-picture understanding of the structure of the solar system is expressed in a simple diagram that resembles a bull's-eye.
  1. The Sun is surrounded in rings by the families of the rocky planets, the asteroid belt, the giant planets, the Kuiper Belt, and the Oort Cloud.
  2. The compositional gradient from rocks near the Sun to ices at the periphery provides clues to how the solar system formed.
  3. In this modern viewpoint, Pluto naturally fits in as part of the Kuiper Belt, whose objects share the same characteristics.

IV. This course is structured in three sections that build knowledge from the ground up.
  A. The first section on foundations of astronomy explains the fundamental observations and concepts that are necessary in creating a scientific picture of how the solar system works.

- **B.** The second section presents the detailed knowledge about the Sun, planets, moons, asteroids, and comets gained since the start of the space-age exploration.
- **C.** The third section delves into the current revolution in our understanding of the outer parts of the solar system and of planetary systems around other stars.
- **D.** Our understanding will progress from how we learn about the universe, to how Earth fits in within the solar system, to how the solar system fits in with other planetary systems.

V. Philosophically, this course is based on simple curiosity about the world around us.
- **A.** Astronomy today is a pursuit in which knowledge is its own reward.
  1. The study of distant stars, nebulae, and galaxies provides little of immediate practical value.
  2. The practical aspects of astronomy were discovered long ago and are taken for granted today.
- **B.** We no longer observe the sky the way we once did.
  1. Electric lights greatly diminish our view of the night sky as well as remove the impetus for tracking the phases of the moon.
  2. Atomic clocks provide ultraprecise time, which hides the fact that observations of the sky still define our system of time.
  3. Global positioning satellites have made the Sun and stars obsolete as navigational tools.
- **C.** Despite modern technology, our innate wonder for astronomy still exists.
  1. The dazzling view of the night sky from a dark mountaintop is a sight not to be missed.
  2. The essence of astronomy is simply to wonder about the sky. Wondering leads to observations, observations recognize patterns, and patterns suggest explanations.

Suggested Readings:

Davies, *Beyond Pluto*.

Feynman, *The Pleasure of Finding Things Out*.

Hakim, *The Story of Science*.

Hey, *Solar System*.

McFadden, Weissman, and Johnson, *Encyclopedia of the Solar System*, chaps. 1, 3.

Sobel, *The Planets*.

Weintraub, *Is Pluto a Planet?* chaps. 1, 11.

Questions to Consider:

1. Which of the historical changes in perspective about the solar system do you feel is the most important, and why?
2. Do you think the 21st-century view will be the last change in perspective that astronomy will have on the solar system?
3. Why are you attracted to astronomy? What do you expect to gain from its study?

Lecture One—Transcript
The 21st-Century Solar System

Welcome. Today we begin a journey in rediscovering our solar system. Now, I say "rediscovering" because everyone is familiar with the basics. We learned about the solar system in elementary school. It's stuff we've known about for hundreds or even thousands of years. The solar system is our backyard in the universe, and we've explored it thoroughly—with telescopes on the ground and in orbit, with manned missions to the Moon and satellites crisscrossing the solar system. We even have an expression for it: There's nothing new under the Sun. This course will show you just how wrong that maxim is.

So, my first goal of the course is to teach you that scientific knowledge is never static. This has always been true, but recently it's become more prominent, and I can sum it up in just one question. Why is Pluto no longer a planet? For 70 years, Pluto was a planet, and then a few years ago, people started talking about kicking Pluto out of the solar system. In 2006, we moved to having 12 planets for about a week, and then Pluto became classified as a dwarf planet. What's the matter? All of a sudden it's too small?

Well, the truth is a little more straightforward because, originally, we thought Pluto was several times larger than Earth, but we now know it to be smaller than our Moon. In fact, it's smaller than seven moons of the solar system. Over the years, we've learned that Pluto is unlike any of the other planets. It's basically a giant ice ball. It's got a very elliptical, elongated orbit. It's got a tilted orbit, and if you try to put it in with the rocky planets or you try to put it in with the gas giant planets it ends up being the ugly duckling of the solar system.

Pluto is so small and so far away that this is one of the best images we can get of it in visible light. This is not an image from a ground-based telescope. This is an image from the Hubble Space Telescope. That's all we get, and that's not what most people expect for a planet.

That brings me to my second goal of this course, and that's to get you to recognize that some of what you think you know about the solar system is actually wrong. Misconceptions about the universe are rampant, and for the solar system, we've got sort of a "been there, done that" hubris that leads to mistakes. We generalize and assume everything follows the same basic rules. We oversimplify, and sometimes the cartoon version replaces the complex reality. We can exaggerate and select facts out in order to make a story or a book or a movie. Too often, our invented reality replaces the scientific view.

For Pluto, the press coverage created a reality distortion field because the reason for Pluto's demotion is not the fickle astronomers who couldn't figure out whether it is a planet or isn't a planet or is a dwarf planet or whatever that really means. And it's not really Pluto itself. The fact is that our knowledge of the solar system has changed. We have discovered a brand new region of the solar system that contains over 1000 new objects.

And so my third goal in this course is to change your perspective, not on the facts of the solar system—you're going to get a ton of new facts about the solar system—but, more importantly, how you organize and understand those facts. In education, we call this a "mental model." It's your framework for discussing things and for understanding new things. And if you have errors in it, these misconceptions can impede the development of new learning. So we want to correct those errors, but we also want to augment or even entirely replace your mental model. And we're going to replace it with something I call the "21st-century solar system."

This is the modern view of the solar system. You see, Pluto hasn't been kicked out of the solar system. Really, the solar system has been reorganized, and Pluto has a new place within it. Now, before describing this new organization, let me just review the previous perspectives we've had on the solar system throughout history.

The first perspective is the Earth-centered solar system, what we call the "geocentric" model of the solar system. This is associated with Ptolemy, who wrote the book *Almagest* in 150 C.E. Now, when we think about the Earth-centered solar system, it sort of gets a bad rap because when we talk about it, we focus on the errors, why it's wrong. We know today that it's not the correct model, but really, it was quite a success because it was the model that first explained the motions of the heavens.

The next version would be the Sun-centered model of the solar system. We call this the "heliocentric" model. This is associated with Nicolaus Copernicus, who published his book *On the Revolutions of*

Heavenly Spheres in the 16th century. But then Copernicus starts to get too much credit because, really, the heliocentric model as presented by Copernicus could not provide better predictions of the positions of objects in the sky than the geocentric model. What the Copernican model really did is it moved from a mathematical description of the motions on the sky to a model that represents reality.

The next model is something I like to call the "gravitational solar system." This has its derivation in the laws of motion, specifically the law of gravity from Isaac Newton. At this point, our perspective of the solar system changed in that we developed a mathematical basis, that there is physics underlying the astronomy, and that we could use that physics to predict motions and predict how the solar system would behave.

The next change in perspective on our solar system really came about with the space age because then we moved from observing the solar system to exploring it. We had manned missions to the Moon. We have satellite missions across the solar system. With the space age, we had developed new technology and new ways of exploring our solar system to develop a rich and detailed knowledge of the objects within it.

Finally, we get to the modern perspective: the 21st-century solar system. This one is still developing because we have got observations of a new region, but we also had discovered other solar systems, planetary systems around other stars. So, not only are we gaining perspective on the structure of our own solar system, we are comparing it to solar systems around other stars and gaining perspective on its structure and its universality.

Now, let's describe this perspective in a little detail. When I first learned about the solar system, the method of teaching it was what I would really refer to as a factoid-based solar system. You know, in our books, we had one chapter on the Sun, one chapter on Mercury, one chapter on Venus, one chapter on Earth, and so on and so on. And really, some of the focus was on memorizing a lot of facts.

So, if we went through that version of teaching the solar system, it would be something like this: This is the Sun. It's a great big ball of hot gas. It's 100 times larger than Earth, and there are gigantic magnetic storms on it that spew fierce radiation at Earth. Then we'd go to Mercury, and Mercury, well, it's cratered like the Moon. It's 450° hot on one side and −200° cold on the other.

Venus. Venus is Earth's hotheaded twin. It's totally covered in clouds, and its surface is so hot that you can melt lead. Earth. Earth would be our planet. It's sort of a water planet, and it has plate tectonics on it that rework its entire surface, but most importantly, Earth is the planet with life. And then to Mars, which is rusty, dusty, and red. It has no air, no water, and subzero temperatures, but for some reason, we still want to go there.

Jupiter. Jupiter is the biggest planet. It's got this Great Red Spot here that's as large as Earth, and this is a storm that has lasted for over 350 years. It's also got four big moons that were seen by Galileo 400 years ago. Saturn. Saturn is the ringed planet. It's got about 60 different moons, and it's got the only moon in the solar system that has an atmosphere. Uranus. Uranus is just a pale bluish-green planet, but it rotates on its side, and that means its poles see 44 years of daylight followed by 44 years of night.

Neptune. Neptune is the blue planet, but it has no water. It's got methane ice clouds that are 200° below zero. It's got this Great Dark Spot that was there in the '80s but somehow was gone in the '90s. Pluto. Well, it's our most distant member, although it was closer than Neptune for two decades. Pluto is smaller than our Moon, and it has a moon about half its size. It's the coldest place in the solar system, −220°.

And then you might get a chapter on the moons. You learn that Earth has one, the giant planets have many. There are about 150 total moons, and they feature some really weird ones, like Mimas, which is a moon that looks like the Death Star from *Star Wars*; or Iapetus, which is a moon that's entirely black on one side and entirely white on the other; or Europa, a moon that has a subsurface ocean and perhaps might harbor life in that ocean.

Finally, you might get a chapter on asteroids and comets, and these would be rocks in space or snowballs in space. And these are on these elongated elliptical orbits, crossing throughout the solar system, smashing into things—and one of them killed the dinosaurs.

So, the way I learned the solar system was focused on memorization, really just gathering together a whole bunch of facts. And what I don't like about it is that it highlights the differences between the objects. You really get little or no feel for the big picture of the solar system. What's the worst thing about this way of teaching the solar system is that it's often presented in an image like this. This is a horrible way to present the solar system—just don't go there.

It's just plain wrong. I mean, the sizes of the planets are sort of done to scale, but the Sun is totally off scale. The spacing between the planets—totally wrong. The orbits are all on a plane? Wrong. These are just characteristic ideas that fundamentally skew your mental model. The most important one is that the planets are all lined up. The planets will be lined up in the solar system once every 3 quadrillion years. That's 600,000 times the current age of our solar system, so you can say that the planets have never been aligned in the entire history of the solar system.

Improving on that model of learning the solar system, we move to comparative planetology—with a space age exploration, which I remember being highlighted, by Viking to Mars and the Voyager missions to the outer planets. We learned to go deeper and see more things and understand more ideas about the objects in the solar system. The basic premise of this is to discuss broad ideas and compare them across the solar system. It's a comparison and contrast, where you highlight the similarities.

An example would be to discuss storms. Here is a picture of Hurricane Isabel, which arose in 2003. It's a very big storm on Earth that formed out in the Atlantic and then swept across the East Coast of the United States. Now, this is an impressive image of a storm, but when you compare it to Jupiter and the Great Red Spot, it can seem rather small, especially when you put the Great Red Spot in context, and put the Earth in for scale. The Great Red Spot is a giant storm on Jupiter that has been swirling for 350 years and is as large as our entire planet.

Doing that comparison allows you to see that what we think is huge here on Earth isn't so huge out in the solar system. Now, you can do many other comparisons. For craters, you can look at Barringer Crater in Arizona and compare it to Mare Orientale on the Moon, or the great ringed Valhalla crater on Callisto. You can look at canyons and compare the Grand Canyon on Earth to Valles Marineris on Mars. Volcanoes: Mount Saint Helens versus Mount Olympus on Mars, versus Jupiter's moon, Io, where an average of seven volcanoes are erupting at any time.

You can look at weather. You've got these hundred-year hurricanes. You've got methane lakes. You've got sulfuric acid rain. These comparisons can continue for rings, moons, magnetic fields, and many other ideas. The messages that comparative planetology sends are that what happens here on Earth happens elsewhere in the solar system. The solar system is understandable. We can model processes here and understand them as they occur out there.

So the progression that you go through is that first you have to establish the facts, then you group these facts into themes, and then these themes can illuminate many different aspects. However, the true big picture is still a little lacking. And that's where the 21st-century solar system comes in.

Instead of talking about the Sun and nine planets, asteroids, and comets, what we do is we group the solar system into six families. These are classes of similar objects, objects that have roughly the same size. Their composition is the same. They're made out of roughly the same type of stuff. Their orbit sizes and their orbit shapes (whether they're circular or elliptical), the inclination of the orbits (whether they're highly tilted or whether they're mostly in the plane)—these are all similar. How many moons do they have? Do they have rings? How strong is their magnetic field?

You look at the characteristics of all these objects and then create families out of them. So, let's go through the solar system one more time and use this modern perspective. Well, there's the Sun. The Sun is a singular object. It is the only star in the solar system. It contains 99.9% of the mass of the solar system, and things both literally and figuratively revolve around the Sun.

Outside of the Sun, in the nearest orbits, we have the rocky planets. And in this image, I will note that I'm showing them to you as they should be seen. Many times, you will see images of these planets that do not show them as they appear with visible light. These are all done to approximately what the human eye would see, especially Venus here. Venus doesn't look too interesting with visible light, so people often show you a radar image or how it looks in ultraviolet light.

So, you can see the rocky planets. They're all made of rock. Their orbits are close in to the Sun. They have few to no moons, and they have relatively thin atmospheres. In the next layer in the solar system is the asteroid belt. These are small to tiny rocks. They contain mostly rock, but they have some ice. The orbits are concentrated between the orbits of Mars and Jupiter, and their orbits are more elliptical and more tilted orbits.

Then outside of that we have the giant planets: Jupiter, Saturn, Uranus, and Neptune. These are very large planets; hence we call them the "giant planets." They have deep gaseous atmospheres. All of them have rings, and all of them have moons. Some of them have very many moons. And they all have very large magnetic fields.

Beyond that is a region called the Kuiper Belt. This is the new region of the solar system. There are over 1000 objects that we have discovered in this region since 1992. These are small to tiny objects. They're mostly made of ice, with some rock. Their orbits, like the asteroids, are more elliptical and more tilted than the planet orbits.

What do they look like? Well, here's an artist's illustration of one. It's an object called Quaoar. No, I'm not making that up. It really is named Quaoar. It was named by a friend of mine for a goddess of a Native American tribe in Southern California. It is basically a giant ice ball, and you can see it. The artist has drawn some craters on it, but you've got to wonder what would happen if this ice ball ventured near the Sun. Well, then it would look like this. It would become a comet. The ices on its surface would melt a little bit, and it would develop a tail.

And that is what the Kuiper Belt is—it's a reservoir of our short-period comets in the solar system. Well, if there's a reservoir of short-period comets, that implies there must be a reservoir of long-period comets, and that is what is shown here in this illustration of the Oort Cloud.

Now, let me note that this illustration is logarithmic. If you look at the axis drawn here, it goes from 1 to 10 to 100 to 1000, so the distances aren't linear. Where the orbits of the planets extend out to about 30 astronomical units (AU), where 1 AU is the distance from Earth to the Sun, the Kuiper Belt extends out to 50 AU, and the Oort Cloud extends out to about 50,000 AU. It's about 1000 times more distant than the Kuiper Belt.

Now, we have no observations yet of the Oort Cloud, but we do have a few tantalizing ideas from observations of an object called Sedna, which seems to exist in the no-man's-land in between the Kuiper Belt and the Oort Cloud.

So, the 21st-century solar system can be summarized with a relatively simple diagram. This is what I call a bull's-eye diagram of the solar system. It was just a simplified diagram I created to just show off the basic organizational structure, but as I've presented it to teachers and to students, I recognize that in its simplicity, it has a lot of power because it shows the families of the solar system in embedded shells.

You have the Sun at the center. Around that, you have the rocky planets. Around that, you have the asteroid belt. Around that, you have the gas giant planets. Around that, you have the Kuiper Belt. And finally, on the outer edges, you have the Oort Cloud. You can see the distribution of the families' objects in the solar system. And when you look at it, you can see that it has a compositional gradient. The rocky stuff, the rocky planets and the asteroid belt, are on the inside. The icy stuff, the Kuiper Belt and the Oort Cloud, are on the outside. In between, you've got the gaseous stuff.

Well, that makes sense because this gives you some clue to the formation of the solar system, where you've got this hot Sun in the center, and you couldn't have the ices forming close to this hot Sun. They had to be way out on the edge. And the rocks, well, they could be in there. So, this is the perspective that changed Pluto—because when you look at this 21st-century solar system, Pluto is best understood as a member of the Kuiper Belt.

I presented some of this material to my daughter's kindergarten class, and at first they weren't too happy because it seems that Pluto is every kid's favorite planet. I don't know, being small, maybe they associate with the smallest planet. But as I presented it, and as I gave them the ideas, they came back to me with a fairy tale. And the moral of their fairy tale was that Pluto has now found his family. It was an ugly duckling story, where Pluto goes from not fitting in to eventually being a member, fitting in with the Kuiper Belt.

Now, this is just one of many other new perspectives I want to present to you in this course. And to do it, we've structured the course into three sections. What we're going to do is sort of build the knowledge of the solar system from the ground up.

In the first section, we're going to establish the foundations of astronomy. We'll explore the historical perspectives. We'll go through the fundamental observations of the sky. We'll lay down the concepts in physics, of gravity, motion, light, and other things. We'll go through astronomical observing. The basic idea is to work into creating scientific models of the solar system. These are all the ideas necessary to develop a scientific picture of how the solar system works.

In the second section, we'll go through the detailed knowledge of the Sun, planets, moons, asteroids, and comets—especially that provided by the space-age solar system. There's been an exponential growth in our knowledge, and there's some really, really cool ideas that I'm going to love to present to you. We have literally volumes of new information, and we'll use that modern approach. We will use comparative planetology to go through lectures on craters, or lectures on weather, or lectures on atmospheres and crosscut the solar system.

In the third section, we will go through the current revolution in our understanding. We'll explore the outer parts of our solar system, the Kuiper Belt and the Oort Cloud, and we will also gauge our understanding of planetary systems around other stars.

So the development as we go through this course will be: first, how we understand the universe; second, how the Earth fits in within our solar system; and third, how our solar system fits in with other planetary systems. The philosophical basis for this course is really just curiosity because the study of astronomy today, the knowledge gained, has to be its own reward. When we look at distant planets, stars, nebulae, and galaxies, we find very little of immediate practical value.

The practical aspects of astronomy—be it timekeeping or the calendars or navigation—well, these were all discovered long ago, and we sort of take them for granted today. We don't observe the sky the way we once did. It used to be very important to keep track of the sky. The full Moon used to be important. If you wanted to do something in the evening, well, knowing when the full Moon occurred would be very good. You could do things in the evening during the full Moon. And this is probably the source of stories about werewolves coming out at the full Moon, or lunatics, where *luna*, the root word, stands for the Moon.

We have electric lights now, and so we no longer need to track the phases of the Moon. And these electric lights also greatly diminish our view of the night sky. As astronomers, we fight something we call "light pollution," and it's really just all the lights of cities going up into the sky and reducing our view of the magnificent heavens.

We have atomic clocks, and they can provide ultraprecise time, so there's no need to tell time by the Sun any more. Now, it's true that the observations of the sky still define our system of time, but it's not something that you and I have to be worried about. We also have global positioning satellites. They can give us ultraprecise information on where we are. I have one in my car. It tells me where my car is to within, you know, a few tens of meters. The Sun and the stars have now become obsolete as navigational tools.

But despite all this modern technology, our innate wonder for astronomy still exists. The view from a dark mountaintop is still awe-inspiring. A few summers ago, I was vacationing in Yosemite National Park, and I wanted to go out and look at the stars. My son was the only one who would come with me late at night, so we went out to the center of the valley and just looked up. Wow, what a sight!

You know, for those of us who spend our lives in suburbia, to get out to a dark site and really see the sky—it's just amazing. And my son, he's getting to be a teenager, and his dad is no longer the coolest guy. He's been learning about astronomy from me since he was, you know, knee high to a grasshopper. So he's heard it and been there and done that, right. Well, underneath that incredible sky, dad was suddenly cool again; astronomer dad was suddenly cool again.

And he had questions, and he wanted to learn, and he wanted to listen. And there's something humbling, you know, very uplifting about looking at the sky. And it always evokes questions. Whenever people learn that I'm an astronomer, and we're out at a dark site, looking at the sky, I get tons of questions. But that's really where I want to begin this course.

I want you to think about standing there and looking at the night sky, and wondering—gazing at the wonder of the night sky—because when you start to wonder about it, that leads you to make observations. And when you make those observations, you'll eventually find patterns in those observations. And the patterns in those observations will lead you to create explanations about those. In that way, you're not only building the mental model, but you're also understanding the observations upon which it is based.

We'll begin building that mental model in our next lecture. There, and throughout the course, I'm going to introduce you to a ton of new images. I'll present incredible new discoveries. We'll float some really cool new ideas, but mostly we'll lead you to new perspectives on how you think about the solar system. I'm really looking forward to it because it's a wonderful story to hear, but it's also a wonderful story to tell.

We'll see you next time.

Lecture Two
Geometry of the Heavens

Scope:

The simplest explanation for the motions of the heavens is the geocentric one: The Sun, Moon, stars, and planets are orbiting around Earth at the center. The scientific challenge is to develop a physical model that can explain and predict these motions, especially the retrograde motions of the planets. Over 500 years of studies, philosophies, and calculations underlie the publishing of Ptolemy's *Almagest* around 150 C.E. The book systematically develops a mathematical explanation based upon the geometry of circles, off-center circles, and circles upon circles. This Earth-centered model provided predictions of sufficient accuracy that it remained the standard for another 1500 years. More generally, the universe was shown to be comprehensible, and heavenly motions could be made transparent by humankind's ingenuity.

Outline

I. The simplest model of the solar system is an Earth-centered one.
   A. The Sun, Moon, stars, and planets appear to orbit around Earth.
   B. A scientific model of the solar system must be able to account for observed motions and predict future positions.
   C. The geocentric solar system devised by the ancients serves as an excellent example of building up a scientific model.
   D. It is a myth that the geocentric model became too complicated and collapsed under the weight of tweaks and inaccuracies.

II. Many astronomical observations can be made by careful observation and judicious application of geometry.
   A. The shape of Earth's shadow during a lunar eclipse argues that Earth must be spherical.
   B. The relative distances to the Moon and Sun can be calculated by triangulation.
   C. A total solar eclipse shows that the ratio of sizes of the Moon and Sun is the same as the ratio of their distances.

  D. The size of the Moon relative to Earth can be determined from careful observations of lunar eclipses.
  E. The size of Earth can be calculated using noontime shadows.
III. The geocentric model in Ptolemy's *Almagest* (150 C.E.) brought together 500 years of astronomical knowledge and is based on the geometry of Euclid's *Elements* (300 B.C.E.).
  A. The geocentric model of the ancients considers the objects visible with the unaided human eye: the Sun, the Moon, Mercury, Venus, Mars, Jupiter, Saturn, and the stars.
    1. Each object moves across the sky at a slightly different speed.
    2. The ordering of object distances was based on the idea that the closest objects showed the fastest motions.
    3. The observed speed of motion can vary considerably over time.
  B. The eccentric circle can explain the motion of the Sun.
    1. If the Sun showed circular motion, the seasons would all be the same length.
    2. The observed lengths of the seasons differ by several days.
    3. The Sun moves faster, relative to the stars, in January than it does in July.
    4. An eccentric circle uses circular motions in which the observer is located away from the center of the circle.
    5. For an eccentric circle, the motion across the sky is fastest when closest, and slowest when farthest.
  C. The epicycle and deferent system can explain the basic motion of the Moon.
    1. Our modern idea of a moon orbiting a planet that is orbiting a star is a natural expression of the epicycle and deferent system.
    2. The deferent circle is the main orbit, and the epicycle circle is the suborbit, whose center moves along the deferent circle.
    3. The moon's motion can be reasonably fit by one epicycle but even better fit by adding a second epicycle to the system.
    4. Using epicycle and deferent, one can also exactly reproduce the motion of an eccentric circle.

- **D.** The equant is like an extension of the idea of an eccentric circle.
  1. The equant uses circular motion in which both the observer and the center of the rotation are displaced equally from the center of the orbit circle.
  2. The equant produces larger variation between the fastest and slowest motion of an object across the sky.
- **E.** The planets were modeled with an equant and epicycle-deferent.
  1. Mercury and Venus are always observed near the Sun, and hence their motion along the deferent was tied to the motion of the Sun.
  2. When Mars, Jupiter, and Saturn are opposite the Sun on the sky, they show short periods of retrograde motion, moving backward across the sky relative to the stars.
  3. Retrograde motion is produced in the geocentric model when the angular speed along the epicycle is significantly faster than the angular speed along the deferent.
  4. To tie retrograde motion to the time of opposition, the motions of Mars, Jupiter, and Saturn along their epicycles were tied to the motion of the Sun.

IV. Though eventually shown to be incorrect, the geocentric model of the solar system was an intellectual triumph.
- **A.** A full scientific model of the solar system was based on three geometrical constructs: eccentric circle, epicycle-deferent, and equant.
- **B.** Ptolemy's work was entitled *Mathematical Systematic Treatise*, but it was later called *Almagest* by Arabic astronomers.
- **C.** The geocentric model provided accurate enough predictions that it remained the standard for 1500 years.
- **D.** The geocentric model showed the basic tenet underlying astronomy: that the universe is understandable, explainable, and predictable.

Suggested Readings:

Bennett, Donahue, Schneider, and Voit, *The Cosmic Perspective*, chap. 3.

Crowe, *Theories of the World.*
Hakim, *The Story of Science.*
Ptolemy, *Almagest.*
Weintraub, *Is Pluto a Planet?* chap. 2.
Wilson, *Astronomy through the Ages*, chap. 3.

Questions to Consider:

1. Before this lecture, what preconceptions about the Earth-centered model did you have? Did any of those ideas turn out to be misconceptions?
2. The Ptolemaic model can be considered simple, as it relies on three geometrical constructs, or complex, as the motions produce intricate patterns. Which position would you support?
3. Is accurate prediction of events enough for a scientific model, or does it have to truthfully represent reality?

Lecture Two—Transcript
Geometry of the Heavens

Hi. In this lecture, we're going to explore creating a scientific model of the solar system, and a scientific model must be based on observations. We're going to keep it simple and use only human-eye observations, the observations you could do without a telescope. And we're going to create the kind of model that you could create yourself. We're going to follow the development of the ancient Greeks as they put together their model of the solar system, and we'll examine the perspective of an Earth-centered universe.

The Earth-centered universe is really the most natural and simplest of models. We call it "geocentric"—*geo* meaning "Earth"—Earth at the center. And if you just observe the sky, it's the model you would come up with. This image here shows a time lapse of the Sun. We see it rise in the east. We see it set in the west. It certainly looks like it's circling around Earth. At night, you can go out and look at the stars—and here's a gorgeous image of stars in the area around the constellation Orion. But if you look in the Poles and you watch the stars over time—you do a time lapse of the stars—you can see that the stars also make circles in the sky all night long.

So, the most natural model is that everything is circling around Earth. So, this would be your first, initial model. The Earth would be here at the center, and all the rest of the solar system would be circling around Earth. Now, to make this scientific, we have to add in some specifics. What are the sizes of the objects? What are their distances? How do they move? The problem with the geocentric model in modern society is not really with the geocentric model, but it's with the mythos that has developed around it.

We have this idea that the geocentric model was way too complicated, that the Ptolemaic system that was developed by the Greeks started simply—and then added in this feature, and added in that feature, and then it needed this feature, it needed that feature, it needs this feature—and it sort of collapsed under its own weight. This is definitely not true. It's a nice story. It makes us feel good. Oh, we don't believe such simplistic stuff any more. But it's a disservice to science education because the geocentric model has a lot to teach us, not only about the solar system but also about how you construct a scientific model.

The geocentric model, and all scientific models, should be judged on whether they can account for observed behavior. Can you explain the observations of the sky? You also need to be able to predict future behavior. The simplicity and accuracy of the model that can explain observations and predict future behavior is really all that's required. One question we'll also examine today: Is the truth of the model really necessary? If the model can explain what we see, does it really matter if it corresponds to reality?

We'll start with the basic ideas. Let's just simplify things. The Earth, the Sun, and the Moon—what can we tell by simple observations? One thing we can tell is that the Earth is round. Now, I know many of you were taught that Columbus showed the Earth was round when he sailed across the ocean. He gets way too much credit. Actually, it was known the Earth was round by the ancient Greeks. Aristotle, in about 400 B.C.E., could offer several different arguments for why the Earth must be round. Aristotle had this thing for spheres, but he had some scientific arguments behind them.

The easiest way to see it is to take a look at a lunar eclipse. A lunar eclipse is when the Moon passes into the shadow of Earth. Here's an image that shows a time lapse, and on the left-hand side you see the full Moon. And as it progresses to the right-hand side, the shadow of Earth slowly creeps across the surface of the Moon. What you notice when you look at this is that the shadow of Earth is curved. If you look at many lunar eclipses and many parts of Earth's shadow, you find that it is always curved. Aristotle reasoned that the only shape that would always cast a curved shadow would be a sphere. And thus he concluded that the Earth was round.

The Moon can also be used to tell us about the relative distances to it and to the Sun. If you watch the Moon over a period of weeks, you see it change its phases. It goes from a crescent Moon, up to a half Moon, to a gibbous Moon, to a full Moon, and then backwards through those same things: gibbous, half, and to crescent. This is caused by the changing of the angles between our direction to the Moon and our direction to the Sun. Basically, the lit part of the Moon varies according to the Moon's orbit around Earth.

If you look at the half Moon and think about it, it actually is a special time because when the Moon is half lit, that means the Sun is perpendicular to the Moon, that there's a 90° angle between our direction to the Moon, and the Moon's direction to the Sun. When you have a right angle, you then get a right triangle, and you can use trigonometry to deduce answers.

Here is a simplistic drawing of the triangle that would be formed when the Moon is at half phase. The Earth, looking directly at the Moon, sees the Moon at half phase. The Sun would be at 90°. In order to measure the sizes of the lengths of this triangle, it suffices simply to measure the angle on the sky between the Moon and the Sun. If you can measure that angle, which is going to be very large, almost 90° itself, but if you can measure that angle, you can then get the scale of this triangle, and you can understand the relative distance to the Moon compared to the distance to the Sun.

Aristarchus was able to do this in about 270 B.C.E., and I note that this was just after Euclid's *Elements,* the textbook of geometry that was published at about 300 B.C.E. He got his angular measure a little bit wrong and estimated that it was about 20 times, that the Sun was 20 times farther away than the Moon. The real answer is it's about 400, but the method, the geometrical method, was sound.

We can get the relative sizes of the Moon and the Sun by looking at solar eclipses. This is an image of a partial solar eclipse, and here, instead of the Earth's shadow going across the Moon, we have the Moon passing between Earth and the Sun. And you can see the limb of the Moon passing in front of the Sun.

It turns out, and it's a very lucky coincidence, that the angular size of the Moon in the sky is about the same as the angular size of the Sun in the sky, and when they pass directly in front of each other, we get a total solar eclipse. This is a spectacular event, and during that time, the entire disc of the Sun is blocked out and you can see this wonderful corona around the Sun, this amazing stuff that you can't see at any other time, at least not with the human eye.

The geometry of the solar eclipse would look like this: Earth is over here. The Moon is blocking the light from the Sun, and we can form triangles again. There are two similar triangles here, the Earth to the diameter of the Moon, and the Earth to the diameter of the Sun. They form similar triangles, and because you can get a total solar eclipse where the Moon just blocks out the Sun, that means that the size ratio of those diameters is the same as their distance ratio. If the Sun is 400 times farther away than the Moon, then that means it's also 400 times bigger than the Moon.

A lunar eclipse can be used to get the relative sizes of the Earth and Moon. If we take a look at this image, it shows you the full sequence of a lunar eclipse. The Moon starts out not being in Earth's shadow. It slowly passes through Earth's shadow and then emerges from Earth's shadow. And if you time the length of the lunar eclipse, you can estimate the size of Earth's shadow, and specifically, you can estimate the size of the Earth's shadow relative to the size of the Moon.

If you take a look at the geometry of it, it's really just the same thing as before: the Sun casting its light, the Earth creating a shadow, and the Moon moving through that shadow. Aristarchus looked at this, and he estimated that the size of Earth's shadow was about three times the size of the Moon. The real answer is that it's about four times the size of the Moon, but the method, again, was sound. And you're able to then tell that Earth is about four times larger than the Moon by examining lunar eclipses.

The last piece you need to complete this basic model is the size of Earth, and you can get this with a remarkable observation, by when the Sun is directly overhead. A man named Eratosthenes was able to do this, and he lived in Alexandria. He noticed that on a certain day in Alexandria you could look down an entire deep well and see the bottom. The light from the Sun would go all the way to the bottom of a deep well. That meant the Sun was directly overhead.

On that same day, in a nearby city named Syene, the Sun was not directly overhead. If you put a vertical rod in the ground, you could measure the shadow that it made. Looking at the geometry of it, you can see that the length of the shadow at Syene would enable you to measure the angle between Alexandria and Syene along the circumference of the Earth. Then, knowing the distance between Alexandria and Syene, you could calculate the full circumference of the Earth. And he was able to do this and get a very accurate estimate for the size of Earth.

Combining these pieces of information, you get the size of Earth, then you get the size of the Moon and the size of the Sun, and from that you get the distances to the Moon and Sun. You end up with basically looking at eclipses and observations of the Moon and Sun, and from that you could have a full model of Earth, Moon, and Sun.

To make this into a more complete scientific model, you'd also want to look at motions, and motions can provide a basic way to order the distances to objects, the Sun, the Moon, and the planets. If we look at the periods of the various objects, the Moon goes through its phases in about 27 days, about a month—it's the origin of month; Mercury, in about three months; Venus goes through its period in about three-quarters of a year; the Sun in 1 year; Mars, about 2 years; Jupiter, 12 years; and Saturn, about 30 years.

Though we don't know at this point, from the human-eye observations, the distances to the planets, we can get the distances to the Moon and the Sun. We can estimate which one is further away than the other from their motions. It's really just a basic idea that something nearby will move across your field of view at a faster speed than something far away. If far away, it will tend to appear that it's moving slower.

So the basic structure, the ordering of the planets, was determined by the periods. And so in the geocentric model, the Earth was at the center and the planets were around it, moving at slower and slower speeds: the Moon, then Mercury, Venus, Sun, Mars, Jupiter, and Saturn. Now, I've drawn them here as circles, and that would be the basic way you would model them, but the motion that they make across the sky is not constant. It's not uniform, circular motion.

The speed varies along its orbit, and we can tell that by looking at the Sun's motion. The Sun's motion can be measured by the seasons, and the seasons are basically spring, summer, fall, and winter. They are measured by the equinoxes and the solstice, the vernal equinox being the start of spring, the autumnal equinox being the start of fall, the summer solstice being the start of summer, and the winter solstice being the start of winter.

As you can see, they occur in March, June, September, December—approximately three months apart. And if you look at that in a diagram of the Sun's motion around Earth, the solstices are opposite each other on the circle, and the equinoxes are opposite each other on the circle, and they divide the circle into four equal parts.

If the Sun were undergoing uniform circular motion, then the length of each season would be about the same, one-quarter of a year, or about 91 days. However, when you look at the measured time for the seasons, you can see that they are not the same. In this table, I've added in the length of the seasons, and spring is 92.8 days, summer is 93.7 days, fall is 89.8 days, and winter is 89 days. The length of the seasons is not the same, and therefore the Sun cannot be going through uniform circular motion.

How are we going to model that? Well, we're going to look at the model created by a man named Ptolemy, and this is a drawing of him. It is called the Ptolemaic model, but that's really giving him a little too much credit. What Ptolemy did is he took about 500 years of Greek astronomical knowledge and organized it and put it into a systematic presentation. He based the geometry of it upon that found in Euclid's *Elements*. And we're going to look at how they solved these ideas of creating motion.

So if we look at the Sun's motion and the real seasons, we can see that summer is much too long for this arc. This arc needs to be longer. And winter is too short for this arc. Winter needs to be shorter. How would you reconcile that? Well, they came up with the idea that we're going to preserve uniform circular motion for the Sun, but we're going to move Earth off center. If you take Earth and shift it off center, and then the solstices and the equinoxes shift with it, you can see that you can create a much longer arc for summer and a much shorter arc for winter. And you can justify the length of the seasons by uniform circular motion, seen by an observer who's not at the center.

Now, this is a little bit exaggerated here, but we have a more accurate model to show it to you. This is the motion of the Sun in an eccentric circle. The observer is located here at O, and the center of the circle is at Z. The green line is the motion of the Sun circling around Z, and the blue line would be the mean Sun, the Sun as if it were orbiting around the observer at O. If I put this in motion, you can see that the green line and the blue line move at different speeds. If I pause it here, when the Sun is closer to the observer, the green line catches up with the blue line. It moves faster and actually gets ahead of it. Then as the Sun gets further away from the observer, the blue line catches up to the green line, and the Sun observed in an eccentric circle moves slower.

This is what the eccentric circle allowed us to do. It allowed for the Sun to move faster when it was closer to the Earth and move slower when it was further away from the Earth. So the eccentric circle could explain the motion of the Sun. What about the motion of the Moon?

Well, it didn't quite work for the motion of the Moon. For that, we needed a different geometrical construct called "epicycle and deferent." These are really cool words, and they're a little bit scary, but they really are a concept you already know. If you think about a star, and a planet orbiting around that star, and then a Moon orbiting around that planet, you already know epicycle-deferent because this construct of one object orbiting another, orbiting a third is exactly what epicycle-deferent is about. We just have to take this diagram and relabel it.

For epicycle-deferent, we have Earth at the center, the deferent circle is the first circle around the Earth, and the epicycle circle is the second circle around the point on the deferent circle. We're doing uniform circular motion, both in the deferent and in the epicycle. This was used by Ptolemy in order to create a basic Moon model, and we have that for you here.

The observer, again, is at O. So C is the point moving around the deferent circle, and M is the point moving around the epicycle circle. You can see that it makes an interesting motion. But if I turn on the trail because when you think about it, epicycle and deferent could produce some very complex motions, but the motion done for the Moon here, as the trail makes its first orbit, you can see it peruses a relatively smooth orbit, relatively smooth here. And you're able to create a nice orbit for the Moon.

Unfortunately, it doesn't explain the full motion of the Moon. A single epicycle really couldn't explain the speeding up and slowing down of the Moon through its orbit that we observed. Instead, he found he needed to use two epicycles. This is the diagram of Ptolemy's full Moon model. It's a little complicated. We have the observer here at O. We have the first deferent circle at Z. We have the epicycle around Z, C, which is also serving as a second deferent circle. And then we finally have the epicycle M moving around the point C.

The strange thing about this model is that the interior epicycle is rotating in the opposite direction of the exterior epicycle, and so the motion is kind of funky. In fact, they refer to it as the "crank mechanism" necessary to get the Moon's motion. Let me turn that on. Now, you can see the motion of the Moon on the exterior epicycle and then it moving around the interior epicycle. It's a little bit complex.

If I turn on the trail, you'll see that, once again, we have a relatively smooth orbit, even though it's got a relatively complex behavior behind it. Okay, so you can see that the Moon was modeled by a double epicycle-deferent because it had a much larger variation in its motion between its slowest point and its fastest point in its motion.

The interesting thing about the epicycle-deferent is that it's actually equivalent to the eccentric circle. You can construct an epicycle-deferent to exactly match the eccentric circle. And I go back to the Sun model, and you'll notice down here I have a slider that I can go to the eccentric version of it, and so this is the eccentric version of the Sun, but I can also take the slider over to the epicycle version of the Sun and produce exactly the same motion.

A properly configured epicycle-deferent, in which the rotation of the epicycle exactly matches the rotation around the deferent, can produce the same motion. And you can see that clearly when I put the slider halfway in between. In that case, you can see both the epicycle and the eccentric circle peruse the same motion for the Sun. So the epicycle-deferent is sort of like a generalization of the eccentric circle.

There is still one more geometrical construct that we need to finish off Ptolemy's model. It's called "equant motion," and it's like an extension of the eccentric circle. This diagram here shows you circular motion, and the time it takes to go between any one of the sections is the same. That's basic circular motion. Then we can think about having that motion with respect to an off-center point.

If we move the center of the circular motion off from the center of the circle, then you can see these top arcs are shorter than these longer arcs. And if it still takes the same amount of time to go between any of these two points, then the planet must be moving slower between T_1 and T_2, for example, than it would be moving between T_4 and T_5 because it has a shorter distance to cover between T_1 and T_2 and a longer distance to cover between T_4 and T_5.

Equant motion extends a little bit further by also moving the observer off center. And so the center of the circle is here in this gray dot, the center of the circular motion is this unfilled dot, and the observer is here in this blue dot—the result of which is the equant motion observed by the observer has a very strong differential from one side to the other. You can see that the distance here between T_1 and T_2 is much, much shorter than the distance between T_4 and T_5.

Again, when the object is closer to the observer, it moves faster. When it's further from the observer, it moves slower. So, we've got the Sun and the Moon, and now we're going to add in the planets. The planets were modeled with the equant motion and the epicycle-deferent. You need the epicycle-deferent because you've got to explain the strange motion of the planets.

The derivation of the word "planets" is "wanderer." They were the objects that wandered across the sky. The Sun and the Moon always moved in the same direction across the sky, and never in the reverse direction. The planets did not.

Mercury and Venus, shown here, would always be found near the Sun. Whenever you see a picture of Mercury or Venus, you're going to see it at twilight, either in the evening or in the morning, because it's always found within a certain angular distance from the Sun: Mercury, about 28°; and Venus, about 46°. And their wandering motion is that they would wander back and forth on the sky, around the point of the Sun. You would never find them opposite the Sun. They would always be over in the same region of the Sun, wandering back and forth.

The other planets—Mars, Jupiter, and Saturn—showed a different kind of wandering motion, something we call retrograde motion. Here's a diagram of the path of Mars in a retrograde motion. It is moving across the sky in the normal direction, the same direction as the Sun and the Moon move across the sky, and then it slows down and stops at what we call a stationary point, and then it backs up and goes backwards relative to the stars, hits another stationary point, stops, and then moves forward again.

This retrograde motion was found to occur when the planets were in the opposite position in the sky of the Sun. If the Sun was over here and Mars was over here, that's the time that Mars would go through retrograde motion. And so with the equant and the epicycle-deferent, you're able to model these motions for Mercury and Venus and for Mars, Jupiter, and Saturn.

This is a model of the motion for Mars, Jupiter, and Saturn, and the observer here is at O. The center of the deference circle is at D. The equant motion offset is E. The center of the motion on the deferent circle is C, and the planet is moving at P. Let me put this in motion for you. As you can see, it basically looks like the basic epicycle-deferent. You can't really tell that the center of the circular motion is at the equant point. It looks pretty much like a standard epicycle-deferent.

The important point for these exterior planets is that when they come in the inside part of their epicycle, they make that little curlicue. This is the time when you would see the retrograde motion. The proper ratio of motion around the epicycle with motion around the deferent produces these little curlicues and produces the retrograde motion on the sky.

It also shows you why Ptolemy is famous. He invented Spirograph about 2000 years before it came out as a kids' toy. Now, this is how they produced the entire model of the solar system in the geocentric model, using these three basic geometric constructs, the eccentric circle, the epicycle and deferent, and equant motion. And so to put it all together, we have a complete model of Ptolemy's solar system.

And so here we look at this, and on the exterior we see the sphere of the stars. Basically, I'm moving it in this. The blue represents Saturn. The purple represents Jupiter. The red represents Mars. And we've got a wonderful slider down here, where we can move in closer because the scale of this is much, much too large to look at in one thing. And we move down. We see the Sun here in blue. We see Venus here in green. Going in further, we see Mercury here in red and the Moon here in blue. And way down deep at the center, we find the Earth.

The person who created this model was cute. He added in Ptolemy's map of the Earth. This is a map produced by Ptolemy as a texture onto the Earth. Let's back out a little bit and put this in motion. So, here you can see the Moon working, with its crank mechanism and its double epicycle-deferent. And we pull out, you see Mercury in red, moving with its epicycle-deferent, and to Venus in green, working with its epicycle-deferent, and the Sun, moving in its eccentric circle.

One thing to notice is that the epicycles for Mercury and Venus—Mercury in red here, and Venus in green here—are always aligned with the Sun. The motion on the epicycle allowed for this bounded elongation of Mercury and Venus, that the planets would be able to go in front of the Sun and behind the Sun, as long as the center of the epicycle circle always pointed toward the Sun. So the model naturally is tied to the Sun, and that explains the bounded elongation.

We move further out. We see Mars in red, Jupiter in purple, and Saturn in blue. Let me speed it up a little bit. And you can see the eccentric motion of the epicycle, the sort of curlicue motion. The other thing you want to notice, if I pause it here, is that the positions of Mars, Jupiter, and Saturn, in their epicycles, are tied exactly to the position of the Sun relative to Earth. That allows the retrograde motion to occur at opposition.

I'll put it back in motion again, and you can see that the motion on their epicycles is tied to the Sun's motion around the Earth. When viewed in this manner, the full Ptolemaic model looks very complex, but the secret is that it is complexity derived from simplicity. Remember that there are just three geometric constructs that underlie all the motions observed in this model, and in that manner, and in many other respects, this should be considered a triumph—not just because you can create a complex model out of simple ideas, but because this was the first time that the motions of the heavens had been scientifically predicted.

And it could not just predict the motions of the planets and the Sun and the Moon—it could also predict other events, like eclipses, which had never been able to be predicted before. Ptolemy did not think this was a real physical model, but rather just a mathematical device for predicting. He entitled his book *Mathematical Systematic Treatise*. It was later called *Almagest* by Arabic astronomers. This geocentric model provided accurate enough predictions that it remained the standard for 1500 years.

The only work that I know of that lasted longer was Euclid's *Elements*, which was a standard textbook of astronomy well through the 19th century. As a matter of fact, the geocentric model is the model you would derive today if you just based your observations on what the human eye could see—you would come up with a model like this. And that shows a basic tenet that underlies astronomy, and all of science: that the universe is understandable, that it's explainable, and that it is predictable. That is the construct of science.

However, this model leaves one question open. Is mathematical prediction enough? A scientific model that can explain the observations is great, but one that actually represents reality would be better. Does a model actually have to represent reality, or is just the mathematical prediction enough? We'll explore that question in our next lecture. Until then, have a great day.

Lecture Three
Truth, Beauty, and Heliocentrism

Scope:

The geocentric model of the solar system provides adequate predictions of the positions of the Sun, Moon, and planets for human-eye astronomy. In 1543, Copernicus presented an alternative hypothesis of a Sun-centered, or heliocentric, solar system. Being based on the same geometrical constructs as its predecessors, the model was not scientifically more compelling. One could argue that it was philosophically more attractive, especially in that it raised the discussion from a calculational device to a physical model of nature. Still, given the state of science and mathematics in the early 1600s, even Galileo could not prove the correctness of the heliocentric model. The fundamental lesson is that the reference frame of Earth and the reference frame of the Sun provide essentially equivalent vantage points from which to calculate the motions in the solar system.

Outline

I. The geocentric model of the solar system was successful for more than 1000 years.
   A. The geometric calculations were based on inventive combinations of circular motions, including off-center circles (eccentric circle, equant) and circles upon circles (epicycle and deferent).
   B. The combinations that could predict the position of a celestial body were not unique.
   C. During the Middle Ages, Arabic astronomers preserved the knowledge of the geocentric model and developed alternative versions.
   D. At the start of the Renaissance, the Aristotelian philosophy of "perfect" heavens still held sway.

II. The heliocentric model is an alternative hypothesis with deep philosophical implications.
   A. Nicolaus Copernicus's heliocentric model switched the places of Earth and the Moon with that of the Sun.

1. His book, *On the Revolutions of Heavenly Spheres*, was published in 1543, the same year he died.
2. Copernicus worked part-time for more than 30 years developing his heliocentric model, but few details of this development work remain.

B. Many myths about the heliocentric model have arisen in popular culture.
1. It is a myth that the heliocentric model was more accurate in its predictions than the geocentric model.
2. It is a myth that the heliocentric model was much simpler than the geocentric model.
3. It is a myth that the center of the universe is the most important place in the geocentric model.
4. It is a myth that Copernicus delayed publishing for fear of reprisal from the church.
5. It is a myth that the heliocentric model was considered a mathematical calculation device and not a representation of reality.
6. It is a myth that astronomers quickly embraced the heliocentric model.
7. It is a myth that Galileo proved the heliocentric model was correct.

C. The heliocentric model provides several advantages over the geocentric model.
1. It makes more sense that the much smaller sphere of Earth moves, rather than the much larger Sun or the sphere of the stars.
2. The heliocentric model provides a natural explanation for why Mercury and Venus are always found near the Sun.
3. The heliocentric model explains why Mars, Jupiter, and Saturn are closest to Earth when they are opposite the Sun on the sky.
4. The heliocentric model provides an explanation of retrograde motion that does not involve any true backward motion.
5. The heliocentric model is a physical model representing nature, not just a mathematical device.

- **D.** The heliocentric model had several problems.
    1. If Earth rotates, a point on the equator would be moving at an absurd speed of about 1000 kilometers per hour (600 mph).
    2. If Earth orbits the Sun, it would be moving at an absurd speed of about 100,000 kilometers per hour (60,000 mph).
    3. If Earth orbits the Sun, the stars should shift positions over the course of a year due to the parallax effect.
    4. The scriptures state in several places that Earth does not move.
- **E.** Both the advantages and disadvantages of the heliocentric model were essentially philosophical, not scientific, in nature.

III. Given the science of the early 1600s, one could not provide irrefutable proof of the heliocentric hypothesis.
- **A.** Several observations showed that the Aristotelian view of perfect and immutable heavens was wrong.
    1. A supernova explosion in 1572 showed that the stars do indeed change.
    2. Mountains on the Moon showed that celestial bodies were not all perfect spheres.
    3. The moons of Jupiter showed that not everything orbited Earth.
    4. The Tychonian model of the solar system reflected this thinking, with the Sun and Moon orbiting Earth while the planets orbited the Sun.
- **B.** In 1615, the Catholic Church reviewed Copernicus's book and decreed that the heliocentric model could only be called a hypothesis.
    1. Galileo, whose work had inspired this review, was cautioned not to hold or defend the ideas of heliocentrism.
    2. Galileo instead took up the challenge of proving the heliocentric model was correct, and he presented his arguments in his book *Dialogue Concerning the Two Chief World Systems* in 1630.

3. For this work, he was tried by the Inquisition, forced to abjure, and placed under house arrest for the remainder of his life.
C. Galileo proposed that the tides were proof that Earth moved.
1. In physics, the combination of two motions is simply the vector sum of the two separate motions.
2. The combination of Earth's rotational and orbital motion produces a faster motion at the point on Earth farthest from the Sun and a slower motion at the point closest to the Sun.
3. Galileo argued that the acceleration and deceleration in speed produces a sloshing of the oceans that creates the tides.
D. Galileo's argument for the heliocentric model was wrong.
1. The tides are an effect seen from the reference frame of the Earth, while the change of speed in his argument is seen only from the reference frame of the Sun.
2. The high and low tides cycle every 12 hours, while his argument would produce a 24-hour cycle.
3. Galileo was aware of the argument that the Moon produced the tides but dismissed it as "ridiculous."

IV. The shift from geocentric to heliocentric is just a transformation of the heavens.
A. All motion is relative to some defined frame of reference.
B. The various solar system models discussed so far can be viewed as essentially equivalent.
1. The geocentric model compared to the heliocentric model is a change of reference frame from Earth to the Sun.
2. The Tychonian model is a rotational transformation of the heliocentric model.
3. Galileo wrote a note in his copy of his book that could be interpreted as recognition that he had not truly proved that Earth moved.
C. In 1979 the Catholic Church reexamined Galileo's case, and in 1992 Pope John Paul II endorsed its findings that Galileo should not have been condemned.

D. Astronomers would not be convinced to switch to the heliocentric model until it could make better predictions for their observing than other models.

Suggested Readings:

Bennett, Donahue, Schneider, and Voit, *The Cosmic Perspective*, chap. 3.

Copernicus, *On the Revolutions of Heavenly Spheres*.

Crowe, *Theories of the World*.

Galilei, *Dialogue Concerning the Two Chief World Systems*.

———, *Siderius Nuncius*.

Gingerich, *The Book Nobody Read*.

Weintraub, *Is Pluto a Planet?* chap. 3.

Wilson, *Astronomy through the Ages*, chap. 5.

Questions to Consider:

1. What role should questions about philosophical beauty and truth play in science?
2. Compare the simplicity and complexity of the Ptolemaic, Copernican, and Tychonian models. Given that all three used similar geometrical constructs, can any one of them be argued as truly scientifically simpler?
3. Can an idea be considered dangerous? What if that idea contradicts long-held beliefs and/or could lead to social unrest?

Lecture Three—Transcript
Truth, Beauty, and Heliocentrism

Philosophers discuss truth and beauty. Scientists discuss hypothesis and theory. What happens in science when you have two competing hypotheses? Then truth and beauty become part of science. And which is more important? And in science, how would you measure the truth or beauty of a hypothesis?

These ideas come into play as we look at the next perspective on the solar system, the heliocentric hypothesis of Copernicus, "heliocentric" meaning Sun-centered, *helio* being a root word for "Sun." You'll notice that I said hypothesis and not fact because I want to examine it in its historical setting. Although we now know that this is the correct model of the solar system, for about 100 years after it was proposed, it was still just a beautiful hypothesis. Even one of science's greatest minds could not prove its truth.

The geocentric model had been successful for more than 1000 years. It was based on three geometric variations on circular motion. If we have circular motion like this here, we would have uniform motion around a circle. The eccentric circle was a variation, where you viewed uniform circular motion, but from a point off center. Then equant motion went a little bit further, and it not only took the observer off center but also placed the center of the circular motion off center. Finally, the third construct was the epicycle and deferent, which was circular motion around circular motion. Together, these built up a wonderful model of the solar system, where Earth was at the center.

In the Middle Ages, however, that science was lost in Europe. The knowledge of the geocentric model was carried on through Arabic astronomers. Not only did they carry it on, they almost revered it. They took Ptolemy's book, and they called it *al majiste*, which comes down to us as *Almagest*. They also improved some of Ptolemy's model to try and make it correspond more to reality.

There were some problems with Ptolemy's model, in particular, because Ptolemy only thought that it was a mathematical model designed to make predictions. For example, if we watch this animation of the Moon as it goes through its phases, you can see that while the Moon goes through its phases, it gets a little bit bigger and a little bit smaller. In Ptolemy's model, the variation of the Moon in distance from the Earth was much, much larger, and would have produced a much larger amount of size change for the Moon.

The Arabic astronomers improved not only the Moon model but also models of the outer planets, because the changes in size were true for them as well. Here's an example of an Arabic model called the *al-Shatir* model, and this was a model of the outer planet. Now, I've put it in motion. You can see it's got several epicycle-deferents. It's a rather complex model, but it was devised to make the size changes that were actually observed for the outer planet match the size changes in the models. They tried to improve it to make it correspond more to reality.

When the Renaissance came to Europe, Ptolemaic models were rediscovered. In 1175, there was a Latin translation of the Arabic translation of the *Almagest*. And in 1515, shortly after the invention of the printing press, was the first printing of the *Almagest*. The other thing that came down in the Renaissance was the Aristotelian philosophy, the idea that the heavens were perfect, and that spheres and circles were the perfect shapes.

There's a strong contrast in beliefs here of these two ideas that came down. Ptolemy believed that this was just a mathematical model—it's just a description of movement in the universe—whereas Aristotle believed it was reality. His model believed that there were crystals and spheres in the heavens that guided the motions of the Sun, the Moon, and the planets.

Into this we bring the heliocentric model. It is basically an alternative hypothesis to the geocentric model at this time. This is the geocentric model. You can see Earth is at the center, with the Moon, Mercury, Venus, the Sun, Mars, Jupiter, and Saturn orbiting around it. The heliocentric model simply switches the place of Earth and the Moon and the Sun. The Sun switches to the center; the Earth and the Moon switch out there.

This is the model embraced by Copernicus, and it was presented in his book *On the Revolutions of Heavenly Spheres*. That was published in 1543. Incidentally, it was published the same year that he died. There is an apocryphal story that says on his deathbed, they brought him a copy of his book and he was able to see his life's grand achievement before he died. We don't really know if that's true.

There's a lot else we don't really know that's true because Copernicus did not keep notebooks. We don't know much about what he did. All we really know is that he worked part-time on his heliocentric model for more than 30 years. In the absence of knowing what did happen with Copernicus, filling in the void, there are many myths that have grown up about the heliocentric model. And so instead of studying what we do know, which is kind of scant, let's take a look at all these myths, which are probably almost as large, or larger, than what we do know.

Myth number one: The heliocentric model is much more accurate than the geocentric model. This is not true because Copernicus did not use any new mathematical techniques. There were no significant advances in either observations or observing tools, and therefore the heliocentric model provided no greater accuracy in predictions.

Myth number two: The heliocentric model is much simpler than the geocentric model. Well, I would argue that the heliocentric model's definitely simpler in terms of philosophy. In science, it's not really simpler because Copernicus used eccentrics and epicycles, the same geometric tools that are in the geocentric model. There were no new mathematical techniques developed for Copernicus's model, so really, from a scientific standpoint, it wasn't any simpler.

Myth number three: The center of the universe is the most important place. The idea here is that the critics of the heliocentric model resisted demoting Earth from its place in the center. Well, if you look at the Aristotelian philosophy, it says, actually, the opposite. Perfection in the universe flows upward. The imperfect flows downward. Instead of the Earth being in a revered place in the center of the universe, the heavens were the revered place, and the Earth was the opposite of that. So the Earth not being at the center, from the Aristotelian point of view, would actually be a promotion.

Myth number four: Copernicus delayed publishing his work for fear of reprisal from the church. This I refer to as the "Galileo myth" because Galileo got in trouble with the church for his heliocentric views, so of course Copernicus had to get in trouble with the church for his heliocentric views. That's totally wrong as well because Cardinal Schönberg and Bishop Giese were two major supporters of Copernicus, and they urged Copernicus to publish. Copernicus even dedicated his book to Pope Paul III, and Pope Paul III accepted the dedication.

Myth number five: Copernicus viewed his system as only a mathematical device, not as a representation of reality. This myth arises because Andreas Osiander inserted an introductory letter, and in that introductory letter, fearing reprisal from the church, he put forth the idea that this was just a mathematical model, that they shouldn't take it too seriously, and please don't get mad at us for doing this. Bishop Giese found this to be horrible, and he attempted to get the letter removed from the publishing of the book.

Myth number six: Astronomers quickly embraced the heliocentric model. Well, this was not true, at least not for about 100 years. The Copernicans were few. There were not many people who were willing to embrace the heliocentric model, simply because it didn't provide any better predictions than the geocentric model. It offered no practical advantages to the astronomers, so by the time of Galileo, there were some Copernicans—folks who believed in the heliocentric model—but not many.

And finally, myth number seven: Galileo proved the heliocentric model was correct. This comes from the fact that Galileo was tried by the Inquisition and got into a lot of trouble for espousing the heliocentric viewpoint, but he did not prove the heliocentric viewpoint. And the reason why he failed illustrates a fundamental physical principle discovered by Galileo himself.

So the heliocentric model, as it would have been studied in the time of Galileo, was viewed as an alternative to the geocentric model. It provided several advantages. If we look at the circles of the heliocentric model, it makes much more sense that the smaller sphere of Earth moves through the universe, instead of the much larger sphere of the Sun. And it certainly makes a lot more sense than the much, much, much, much larger sphere of the stars orbiting around Earth once per day. Smaller things moving makes much more sense than larger things moving.

It also provides a natural explanation for the bounded elongations of Mercury and Venus. As you can see in this diagram, they are interior planets. They're in between Earth and the Sun. And therefore they could never appear on the opposite side of Earth from the Sun. They would always appear interior and towards the direction of the Sun, which is exactly what we observe for Mercury and Venus.

For the exterior planets—Mars, Jupiter, and Saturn—that would explain why they appear largest and are closest to Earth when they are opposite the Sun, because when they are near the Sun in the sky, they are far away, and when they are opposite the Sun, when the Sun is on one side and they're on the other side, they are passing closest to Earth.

It also provides a natural explanation for the retrograde motion of these outer planets. In this diagram, the blue dot is meant to represent Earth, and the red dot is meant to represent an outer planet. Take Mars, for example. As Earth in its inner orbit catches up and passes by the outer planet, the outer planet appears to move forward, across the plane of the stars, then backward, and then forward again. But it does not involve any true backward motion. It's just the perspective of an inner planet overtaking the outer planet because it's moving on a faster inner track.

The final advantage of the heliocentric model is philosophical, that it's a physical model, representing nature, and is not meant to be just a mathematical device. Although the philosophical arguments don't carry too much weight in astronomy, if all things are equal, we'd like to have both truth and beauty.

Now, everything was not wonderful for the heliocentric model. It also had several problems. If I take this globe and I orbit it, I rotate it, if the Earth is spinning once per day, a point on the equator has to go all the way around and back in 24 hours. To do that, it has to be moving at an absolutely absurd speed. You would have to be moving at 1000 kilometers per hour. That doesn't make much sense. Furthermore, if we take this Earth and put it in orbit around the Sun, well, then the entire Earth has to be moving at an even more absurd speed. It has to be moving at 100,000 kilometers per hour. Come on, that doesn't make much sense to move Earth that fast through the solar system.

Also, if Earth is orbiting the Sun, we should see a parallax effect on the stars. That's illustrated by this diagram, that if the Earth is on the near part of its orbit, then the angle to the stars would be larger, and when it's on its far part of its orbit, the angle to the stars would be smaller. The stars should appear to pull away and come closer, pull away and come closer as Earth moves round its orbit. This is what we would call a parallax effect, and it's not observed for the stars.

Now, there is a simple argument that says: Oh, this shouldn't happen. That's just to put the stars immensely far away, but that doesn't make any sense because then you would have this huge empty space beyond Saturn and before the sphere of the stars. And that was considered absurd in this day.

Finally, the most damning argument for the heliocentric model is that the scriptures state in several places that Earth does not move. This doesn't sound like a huge problem to a scientist today, but it was a big problem back then, when the church dominated not only your daily life but also your intellectual career.

All of these arguments, both for the heliocentric model and against the heliocentric model, are essentially philosophical. Some of them have a scientific basis, but they are simply the idea of "Which explanation do you prefer?" Do you prefer to have Earth moving at this incredible speed of 100,000 kilometers per hour, or would you rather have the Sun orbiting around? Make your choice. Which is a simpler argument to you?

The thing in terms of science is that the predictions are equivalent. So, you were an astronomer at the time, what would you think? Would you take an established working model that's been providing predictions your whole life, or would you go with a more aesthetic alternative? How would you choose?

Enter this man, Galileo, the first telescopic astronomer. And his observations with the telescope helped disprove the Aristotelian view. The Aristotelian view came under attack in the era just shortly before Galileo. The idea that the heavens were perfect and immutable could demonstrably be shown to be wrong. In 1572, they observed a supernova explosion, the appearance of a new star that then faded away. That showed that the stars do indeed change.

Galileo's telescope revealed that the Moon had mountains, and it showed that not all bodies in the heavens are perfect spheres. Furthermore, Galileo found the moons of Jupiter orbited around Jupiter, and that showed that not everything orbited around Earth. This break with Aristotle came about and was reflected in a new model of the solar system called the Tychonian model.

If this is the heliocentric model, in which everything orbits the Sun, the Tychonian model would actually be a compromise. It would take the Earth and make it stationary, and have the Moon and the Sun orbiting around Earth, and then it would take all the other planets—Mercury, Venus, Mars, Jupiter, and Saturn—and have them orbit the Sun. It seems like a rather complex idea, but it had quite a lot of support at the time of Galileo.

So, here's where I come to a point, what I call "Galileo's dilemma." Galileo found that he could prove Aristotle wrong, but given the science of the early 1600s, could he provide proof of the heliocentric hypothesis? This question came up in 1615, when the Catholic Church reviewed Copernicus's book. Even though Pope Paul III had signed off on the book 60 years earlier, it came up for review because of Galileo's work. And they decreed that the heliocentric model could only be referred to as a hypothesis. Galileo was particularly cautioned not to hold or defend the ideas of heliocentrism.

However, there was a loophole. There was an implied challenge that if Galileo could provide irrefutable proof of the heliocentric hypothesis, the matter could be reopened. Galileo couldn't resist, and he took up the challenge of proving the heliocentric model. He presented his arguments in 1630 in his book *Dialogue Concerning the Two Chief World Systems*. For this book, he was tried by the Inquisition, forced to abjure, and placed under house arrest for the remainder of his life.

What was his crime? Well, his crime was providing a wrong proof of the heliocentric model. His idea was that the Earth's motion produces the advance and recession of the tides. As he put it, the ocean, because of its fluid nature, is the only thing that can indicate the motion of the Earth. If Earth were stationary, the tides could not occur.

In order to understand that argument, we first have to go through something called the Galilean transformation, and that's simply the idea that in physics, the sum of two motions is a vector sum. If you have two different motions operating, you can combine them with a vector sum. And to see this, let's take a look at the idea of a train.

So, suppose you have a train conductor walking down the aisle of a train. If you're sitting in a seat, you see him walk past you. Say he's moving at about 2 kilometers per hour. However, if you are outside the train, you see the train moving down the track at, say, 30 kilometers per hour. From your perspective outside the train, you add together the 30 kilometers per hour of the train motion, and the 2 kilometers per hour of the conductor walking, and you see his total motion as 32 kilometers per hour.

Which one is correct? The answer is that they both are. It just depends upon the frame of reference. Measurements depend upon how you are moving with respect to what you are measuring. Galileo applied this to his argument about the tides. He argued that Earth's rotation and orbit combined to produce the tides. And the argument looks something like this.

At the antisolar point, the point farthest from the Sun, Earth's rotational motion and Earth's orbital motion combine to produce a faster motion. So far, on the opposite side of the Sun, you're moving at a faster speed. At the subsolar point, the rotational motion is going against the orbital motion, and the two subtract, and you get a slower motion.

His argument was that that acceleration on the far side and deceleration towards the near side created a sloshing of the ocean that created the tides. Wonderful argument. It's just plain wrong is the problem. The problem with this is that the tides are, in effect, seen from the reference frame of Earth, while the change in speed in his argument is only seen from the reference frame of the Sun. He is using a Sun reference frame argument to explain an Earth reference frame effect.

Furthermore, the high and low tides that we see on Earth cycle in about 12 hours. We get two high tides and two low tides every day. In Galileo's argument, you should only get one high tide and one low tide per day. The tides also are observed to occur relative to the sublunar and antilunar points, the points closest to and farthest from where the Moon is at the time.

What's even more intriguing is that Galileo was aware of the correct argument that the Moon produces the tides. In his book, he has the simpleton Simplicio make this argument. The Moon, wandering through the sky, attracts and draws up toward itself a heap of water, which goes along following it so that the high sea is always in that part that lies under the Moon. That's basically a correct argument. However, the wise man Sagredo in Galileo's *Dialogue* dismisses it as "those ridiculous older opinions."

Galileo's proof failed because the shift from geocentric to heliocentric is just a transformation of the heavens. Motion is defined relative to some frame of reference, and the various solar system models are essentially equivalent because they are all just changes of reference frame. To go from geocentric to heliocentric, you're changing from the reference frame of Earth to the reference frame of the Sun.

If you look at the Tychonian model, that's just a rotational transformation of the heliocentric model, taking the heliocentric model and putting it in a spin about Earth. To see that, let's take a look at this animation of the Tychonian model. When you look at this, it looks somewhat like a heliocentric model. It looks like the Sun is at the center. But you'll note that there is a circle going through the Sun, and actually the Earth here is at the center, and the Earth is fixed. If I put it in motion, you can see that the Sun is orbiting around the Earth, but Mercury, Venus, Mars, Jupiter, and Saturn are orbiting around the Sun. This is the Tychonian sort of compromise model.

If I speed it up, you can see that you start to get sort of seasick because it's like a wheel that's out of balance and is wobbling around. You see that same sort of curlicue motion that's produced by the epicycle-deferent in the Ptolemaic model. What we really have here is the Sun working on a deferent circle, and the planets on five epicycles around it. This model was due to Tycho Brahe, and he was one of the greatest pre-telescopic observers. He did his observations in a period from about 1570 to 1600.

He developed this model because he saw the beauty of the heliocentric model and he appreciated the explanations, the natural explanations, for the bounded elongations of Mercury and Venus, and for the retrograde motion of Mars, Jupiter, and Saturn. Yet he could not accept the idea that Earth was in motion. To him, it was still absurd that Earth should be spinning or moving around the Sun.

In the end, all three of the models, the Ptolemaic model, the Tychonian model, and the Copernican model, were scientifically equivalent because they all produced the same observations. Not one of them could predict the motions of the planets any better than any other one.

And really, I think Galileo might have known this because in his own copy of the *Dialogue*, he wrote this quote:

> Take note, theologians, that in your desire to make matters of faith out of propositions relating to the fixity of Sun and Earth, you run the risk of eventually having to condemn as heretics those who would declare the Earth to stand still and the Sun to change position—eventually, I say, at such a time as it might be physically or logically proved that the Earth moves and the Sun stands still.

In a note in his own copy, he was sort of admitting that he had not truly proved his case. The mistake that he felt the church had made was in calling the heliocentric model heretical. And it wasn't until 1992 that Pope John Paul II produced a statement admitting that the church had erred in condemning Galileo.

The philosophical arguments in favor of the heliocentric hypothesis are strong. They are powerful in that it makes much more sense to have everything orbiting around the Sun, but in science, they were not enough. In science, in order to prove the heliocentric hypothesis, you would need better predictions, better motions of the heavens. And this would not come from better arguments, as produced by Galileo, but instead would come from better data, as produced by Tycho Brahe and then analyzed later by Johannes Kepler.

And we'll explore that in much more detail in our next lecture. Until then, have a great day.

Lecture Four
Deducing the Laws of Motion

Scope:

The assumption that the heavens could only be composed of circular motions is a strong limitation of the geocentric and early heliocentric models of the solar system. Generalizing the geometry from circle to ellipse allowed Johannes Kepler to find a better fit to the data. Instead of forcing models to fit assumptions about nature, astronomers look for patterns in the data from which to deduce the nature of the universe. Kepler's three empirical laws not only better predict the motions of the planets but also provide insight into the force behind those motions. Newton's law of universal gravitation is the mathematical expression of that force, and it shows that Kepler's work can be subsumed into an underlying physical principal. More than that, Newton's work embodies the perspective that the same physics that governs motions on Earth also controls motions in the universe. The capability of generalization from local observation to global application is a fundamental precept of all modern science.

Outline

I. In constructing a model of how the solar system works, astronomers search for patterns in the observations that can be generalized into laws of nature.
   A. The geocentric and heliocentric models show that there is more than one way to fit observations.
   B. The fundamental nature of the universe espoused in these models was the philosophy, dating back to Aristotle, that the heavens are perfect and composed of perfect circles.
   C. In the scientific method, the nature of the universe should not be an assumption of the model, but rather derived from the model.

II. Johannes Kepler deduced three empirical laws of planetary motion.
   A. Kepler searched for patterns in planetary data.
   　　1. Kepler succeeded Tycho Brahe in 1601 and gained access to Brahe's precise observations.

2. He spent the better part of a decade engaged in a "war with Mars," trying a wide variety of heliocentric orbit shapes to fit Brahe's observations.
3. Much of Kepler's work would now be written off as numerology, but his quest for order is what led him to discoveries.
B. The orbits of the planets are ellipses with the Sun at one focus.
1. An ellipse is a generalization of a circle that allows for different lengths of the x- and y-axes.
2. From any point on an ellipse, the sum of the distances to the foci is constant.
3. One half of the major axis, the semi-major axis, is equal to the average distance from the Sun.
4. The eccentricity of an ellipse is a measure of the elongation of the ellipse. An ellipse with eccentricity of zero is a circle, while an eccentricity of one means an infinitely stretched ellipse (a line).
5. The eccentricity of planet orbits is generally small, with even the more elongated ones just looking like off-center circles.
C. The orbits of the planets around the Sun sweep out equal areas in equal times.
1. In geocentric models, the eccentric circle and equant were used, in part, to have a planet move faster when closer and slower when farther away.
2. In Kepler's elliptical-orbit heliocentric model, a planet moves faster when it is closer to the Sun and slower when it is farther from the Sun.
3. Circular motion dictates equal angles in equal time, while elliptical motion creates equal areas in equal time.
D. The square of the period of a planet's orbit is equal to the cube of its semi-major axis.
1. Orbit characteristics are measured relative to Earth: Orbit period is measured in years, and semi-major axis is measured in astronomical units.
2. One astronomical unit is the average distance from Earth to the Sun, about 150 million kilometers (93 million mi.).

3. Plotting one characteristic versus another can make patterns easy to spot.
4. Logarithms are a useful tool for finding power law relationships.
5. Mathematically, Kepler's third law is expressed as $P^2 = a^3$.

E. Kepler's laws helped pave the way for the heliocentric solar system.
1. The tables of planet positions based on the geocentric model would lose their accuracy over decades and centuries.
2. The *Rudolphine Tables*, published by Kepler in 1627, were significantly more accurate than any previous planetary tables.
3. Kepler's work embraced and improved the heliocentric model; he actually wrote a three-volume work entitled *Epitome of Copernican Astronomy*.
4. It is interesting to note that Kepler was doing his work at the same time as Galileo but did not incur the ire of the church.

III. Isaac Newton developed the universal law of gravity.
A. Newton's big idea was to connect gravity on Earth's surface to the motions of the Moon and planets.
1. The idea of an apple falling from a tree inspired him to imagine if the Moon were governed by the same force.
2. Newton devised a thought experiment in which he reasoned that if a cannon could shoot a cannonball sufficiently fast, the cannonball would go into orbit around Earth.
3. Gravity is a centripetal force that creates an orbit by continually pulling the path toward the center.
4. The force of gravity between two objects is proportional to the masses of the objects and inversely proportional to the square of the distance between them: $F = GMm/r^2$.
5. Newton's work on gravity developed into his famous three laws of motion.

B. All of Kepler's empirical laws can be derived from Newton's mathematical law of gravity.
1. The force of gravity follows an inverse square law.

2. The general orbit shape for an inverse square force is an ellipse.
   3. The force is stronger when closer to the Sun and weaker when farther from the Sun, leading to the observed variations in motion.
   4. Larger orbits have both longer lengths and slower speeds, leading to the relationship between period and semi-major axis.
  C. Newton's achievement created a physics foundation for astronomical motions.
IV. Science is a search for order in the universe.
  A. Much of science can be described as a three-step process: collecting data, searching for patterns in the data, and explaining why such patterns occur.
  B. The work of Kepler and Newton confirmed that the heliocentric model was the correct one for the solar system.
   1. Kepler's three laws are empirical relationships that describe patterns in the data.
   2. Newton's law of gravity provides the explanation of the patterns.
  C. Our perspective on the motions of the heavens shifted to a firm mathematical basis: the gravitational solar system.

Suggested Readings:

Bennett, Donahue, Schneider, and Voit, *The Cosmic Perspective*, chap. 4.

Berlinski, *Newton's Gift*.

Ferguson, *Tycho & Kepler*.

Shu, *The Physical Universe*, chaps. 3, 18.

Voelkel, *The Composition of Kepler's "Astronomia nova."*

Weintraub, *Is Pluto a Planet?* chaps. 3–5.

Wilson, *Astronomy through the Ages*, chap. 5.

Questions to Consider:

1. Like Brahe and Kepler, astronomers often specialize in either observation or mathematical theory. What qualities do you think lead someone into each specialty? Which would you favor?

2. Kepler openly embraced the Copernican hypothesis while working in Prague. How might Galileo's life have been changed had he not been working in Italy?
3. Can a scientific model ever be said to be more than just a mathematical approximation of reality? Even if it fits all the currently available data, might there be some future data that it will not explain adequately?

Lecture Four—Transcript
Deducing the Laws of Motion

Have you ever played a game when you didn't know what the rules were? Before you can try to win that game, your first effort is to figure out what the rules are. You would try various actions and see what happens. Hopefully you would find patterns in responses, and then you would develop a working model of what the rules of the game were. But you would never know if the rules were exactly right, or if you had a complete set of the rules. You would only know those rules for the situations you had encountered.

Deducing the laws of nature is exactly like playing a game, but in this game, figuring out the rules is the whole point of the game. We make observations, we look for patterns, and we discover what we think are the rules. Our rule book is called science, but we never get to peek at nature's rule book, so we don't know if we've got the correct rules or even if we've got a complete set of rules.

For the solar system, we observe the motions of the Sun, Moon, and the planets. We deduce the patterns that they follow, and we create a model to fit those patterns. As we've shown in previous lectures, both the Earth-centered model and the Sun-centered model can fit the data. Different sets of rules can be found to fit the observations.

However, both models share a common problem. They assume motion based on circles. This can be traced back to Aristotle's notion of a perfect heavens, and circles and spheres as the perfect figures. When we look at nature, we should actually not assume things in the model, but rather the nature should be pulled out of the model. It should be derived from the model. So we need to look for a more general set of rules.

The man who would discover these rules is this man, Johannes Kepler. He found more general rules than the ones we have so far assumed. In 1600, he was hired as an assistant to Tycho Brahe. Tycho Brahe was at the time the imperial mathematician to the Emperor Rudolf II in Prague, and he was the greatest pre-telescopic observer. However, Brahe died one year later. That was very fortunate for Kepler because he succeeded him in the position and he gained access to Brahe's observations.

He used those observations to search for new patterns in the solar system. You see, Kepler was a very strong advocate of the heliocentric model. In 1596, four years before he started working with Tycho, he published a book called *Mysterium Cosmographicum*, and in it he proposed a different model, a heliocentric model that looks something like this.

It was based upon the regular solids and spheres. The spheres represented the orbits of the planets, and the regular solids were the spaces between them. And since there were five regular solids, he justified there being six planetary spheres in a heliocentric model—Earth, of course, being one of the planets. So we have six planets.

But Kepler came up with a real serious problem when he looked at the heliocentric model. He found that he could not fit Mars's orbit. He described it as a "war with Mars," and it lasted for about a decade. As he looked for these patterns in Tycho Brahe's data, he just couldn't fit the orbit of Mars. And Brahe's data was invaluable because most observers would look at Mars only near its opposition, when it was on the opposite side from the Sun, and when it was going through retrograde motion. But Brahe was different than most observers, and he observed Mars and other planets throughout their entire orbit, and that gave Kepler a much deeper range of results from which to compare his ideas.

Today, most of what Kepler looked for, and most of what Kepler found, we would consider numerology. I guess I would describe it as an overzealous quest for order. But he did find generalized patterns, and these are the three empirical laws, Kepler's three laws of planetary orbit.

Kepler's first law is that the orbits of the planets are ellipses, with the Sun at one focus. Now, an ellipse is just a generalization of a circle. We all know what a circle looks like. It's round, right? Well, it has axes that are the same length. The length of the x-axis in this diagram is the same as the length of the y-axis, and it's labeled in this diagram "r." The equation for a circle would be $x^2 = y^2 = r^2$.

An ellipse is just a generalization of that, and it allows you to have different lengths for the x- and y-axes. So this is a diagram of an ellipse, and here a, which is half of the x-axis, called the semi-major axis, and b is half of the y-axis, called the semi-minor axis, and they are different. They're allowed to be different, and that leads to the equation of an ellipse being $x^2/a^2 + y^2/b^2 = 1$. If a and b were both equal to r (if we had a circle), you could multiply through by r^2 and you get the same equation as that for a circle.

A circle has only one special point, the center, and every point on a circle is equidistant from the center. For the ellipse, it has two special points. These are called the focus points, or foci. And you can find them by creating a right triangle with the y-axis and putting the hypotenuse of length of the semi-major axis. And you get your focus point one over here on the right, and focus point two over here on the left. The reason these focus points are special is that the sum of the distances from any point on the ellipse to both of the foci is constant.

I find that a little hard to imagine in my head, but it's relatively easy to see if we do a small demonstration. I have here two pushpins and a string tied between them so that the distance along the string, if I stretch it taut from one focus along the string to the other focus, is going to be constant. If I stretch out the string and draw the figure, I get an ellipse. That's an easy way to draw an ellipse, and you get a much nicer ellipse than I could draw freehand, but you can also see that the sum of the distances is always the same in this shape.

The last thing about an ellipse is the measure of the elongation of the ellipse. We call this the eccentricity. And mathematically it's just the ratio of the axes, and if it is zero, that means if you have a circle—that the two axes are equal. And if you have one, if a/b stretches to infinity, then you have an infinitely stretched ellipse, and that's a line.

So, an ellipse is basically just a generalization of a circle that starts from one extreme, being a circle, going out to the other extreme, being a line. For the planets' orbits, a planet has an elliptical orbit, with the Sun at one focus point. Here it is in this diagram. You can see the planet going around on an elliptical orbit. One focus is not occupied. The Sun occupies the other focus. This means that the average distance to the Sun is equal to the semi-major axis.

Now, the eccentricities we generally draw when we draw ellipses are much, much, much too large for the planetary orbits. I have here a diagram of the inner planets' orbits—Mercury, Venus, Earth, and Mars—and you can see that Venus and Earth look pretty much like circles. Their orbits look pretty much like circles, and in fact they have very small eccentricities.

For Mercury, you can see that the Sun is definitely not at the center of Mercury's orbit. It actually has the largest eccentricity, an eccentricity of about 0.2. But if you look at Mercury, it still just sort of looks like it's a circle, but just with the Sun off center.

You can compare Mars's orbit to Earth's orbit, and you can definitely see that they're a lot closer on one side and a lot further away on the other side. Mars has an eccentricity in between that of Earth and that of Mercury, an eccentricity of about 0.1. It actually looks like it has a stronger eccentricity in this diagram than Mercury, but trust me when I tell you that Mercury has a significantly stronger eccentricity, simply because it's larger and you have something to compare it against.

So, that's Kepler's first law, that the planets travel on ellipses. His second law is that the orbits of the planets around the Sun sweep out equal areas in equal time. In previous models, we had gone from circular motion, which looks like this, where you move through equal areas in equal time—but everything is centered, so you're moving around the same amount and the same size area. Everything is all symmetric.

We then moved, in the Ptolemaic model, to equant motion, which changed this relationship, and the planet had to move slower along one region of its orbit and faster along another region of the orbit. But you can see that the segments that it swept out in the same amount of time had very different shapes.

For Kepler's second law, the idea is the same: that you're going to have the planet move faster when it's closer and slower when it's further away. Kepler's second law looks something like this. When the planet is closer to the Sun, it's moving faster and therefore sweeps out a longer arc, but since it's closer to the Sun, it only has a certain amount of area. As it is further away from the Sun, it moves much slower. It goes through a much smaller amount of arc, but because it's much, much further away, it still sweeps out the same amount of area.

It's very similar to the idea of the equant motion. Matter of fact, if I look at an animation, I can show you that the equant motion and the Keplerian motion are relatively identical. So, in this animation, let's set the eccentricity for the orbit to 0.6, which is much, much larger than a planet orbit, but you can see that the shape of the orbit is distinctly different, and the blue line, which represents the Keplerian motion, and the green line, which represents equant motion, go at very different speeds, especially as they get away from the poles. And you can see how they change and get away and are very strongly different.

If I pause it and set the eccentricity down to something small, which is what the planets have—let's put it around 0.2—and then I put it into motion, you can see two things: that the orbit shape is very close between the ellipse and the circle, and the equant motion is very close to elliptical motion. This is one reason why the geocentric model works so well. The equant motion closely approximated what Kepler would find for the elliptical motion.

Matter of fact, when I look at this I wonder, how did Kepler find discrepancies? I mean, they look so much alike, how could he possibly have found these discrepancies? It just shows you the detail to which he was calculating, and the detail to which Tycho Brahe had observed.

So that's Kepler's second law, equal areas in equal time. Kepler's third law is that the square of the period of a planet's orbit is equal to the cube of its semi-major axis. Let's take a look at the periods of the planets' orbits. In this table, in the first column, the period is measured in years, and it goes from Mercury, at about three months, on up to Saturn, at about 30 years. In the second column, we have the semi-major axis, and that's measured in astronomical units. One astronomical unit is the average distance from Earth to the Sun. It's about 150 million kilometers, or 93 million miles. And you know now, from having looked at ellipses, that that is equal to Earth's semi-major axis.

If we plot these two columns of figures, period versus semi-major axis, we get a graph that looks like this. And you can see that as the period gets larger, the semi-major axis gets larger. And if you look over here on the left, it looks almost linear at the start, but then—no, it doesn't stay linear.

There is a relationship between these two, but it doesn't show up when we plot it in a simple linear plot. If instead we plot it on a logarithmic graph, one can see the power law relationship. A logarithmic graph, such as this one here, is one where the distances along the graph are multiplicative, not additive. In a linear graph, the same distance represents the same amount all the way across the graph. In a logarithmic graph, the same distance represents a multiplication by the same amount.

So for example, if you look at the plot, you go from 0.1 to 1, which is a factor of 10, in a certain distance, and in that same distance, you also then go from 1 to 10. That is also true for any other factor—going from 1 to 2 is the same distance as going from 2 to 4. The point is that in a logarithmic plot, you can see power law relationships. If there's a linear relationship in a logarithmic plot, it means it's a power law. And you can see there is definitely a line through these points in this plot, and I'll draw it here for you.

The slope of this line gives the relationship between period and semi-major axis, and the slope of the line is 2/3, so the mathematical version of Kepler's third law is that it's $P^2 = a^3$. Period squared is equal to semi-major axis cubed. If we put that into a table, you can see that the values of P^2 in the first column, and compare them to values a^3 in the second column, and that the figures match very nicely.

You will note that I did not hide the round-off error, that if you look at Mars, 3.53 versus 3.51. That's just small errors in the numbers that I used. If I had used fully accurate numbers, that round-off error would go away.

The law also shows that the farther-out planets also travel slower than the closer-in planets. Now, you might think that, "Oh well." What this really expresses is that the farther-out planets have longer orbits, and so they could be moving at the same speed, but of course they're going to take longer to traverse their longer orbits.

If the speed were constant, then the period would be proportional to the circumference. And the circumference is just proportional to the radius of the orbit. Then you would get P is proportional to a. Instead, you get P is proportional to $a^{3/2}$. So that means not only do the outer planets have longer orbits, but they also have slower speeds.

Kepler's three laws helped pave the way for the heliocentric solar system. And not in just expressing the three laws as some empirical relationships about the planets, but because the table of planet positions that he created, based upon these three laws, were better in their accuracy than the geocentric model. The planet positions based upon the geocentric model would lose their accuracy over the course of decades or centuries and needed to be recalculated.

In 1627, Kepler produced the *Rudolphine Tables*, and they were significantly more accurate than any previous planetary tables. To an astronomer, you're saying: At last, finally there's a practical advantage to the heliocentric model. This is what you're looking for. While the philosophical arguments are nice, they aren't compelling. The better predictions are enough to turn the tide in science.

Kepler embraced and improved the heliocentric model, and he actually wrote a three-volume work called the *Epitome of Copernican Astronomy*. He was a strong Copernican at a time when, we know from Galileo's efforts, it was not a good time to be a Copernican. He died in 1630, the same year that Galileo published his *Dialogue Concerning the Two Chief World Systems*, but he did not incur the ire of the church, probably because Galileo was in Italy and Kepler was in northern Europe.

Kepler developed the observational basis for the heliocentric model, but Isaac Newton would then develop the mathematical basis. The big insight inside Newton's universal law of gravity, is to connect gravity on Earth's surface to the motion of the Moon, the Sun, and the planets.

Now, there's a myth out there that says that Newton was sitting underneath an apple tree, an apple falls and bonks him on the head, and he thinks: Oh, hey, gravity. No, that's not really quite how it happened. Instead, the idea of an apple falling from a tree inspired him to imagine if the Moon were governed by the same force as that apple.

I have a nice quote here from John Conduitt, Newton's assistant at the Royal Mint:

> In the year 1666 he retired again from Cambridge to his mother in Lincolnshire. Whilst he was pensively meandering in a garden it came into his thought that the power of gravity (which brought an apple from a tree to the ground) was not limited to a certain distance from Earth, but that this power must extend much further than was usually thought. Why not as high as the Moon said he to himself and if so, that must influence her motion and perhaps retain her in her orbit,

whereupon he fell a calculating what would be the effect of that supposition.

So the role of the apple was to inspire him to think about gravity in a much more general way. Could gravity apply to the heavens as well as applying to Earth? How do you then go from gravity on Earth to orbits? Well, Newton came up with a thought experiment. A thought experiment is a kind of experiment that you can do only in your mind. You can think it through, and think through all the ramifications, but you don't actually have to perform it, so they're very nice experiments.

So, this is an illustration that shows the idea behind Newton's thought experiment. What if you had a cannon that was really, really powerful on an airless world? Normally, a cannon would shoot the cannonball, and it falls to the ground. But on an airless world and with a really, really strong cannon, it could fall further and further away, and eventually it might fall and miss the Earth. What would that be then?

Newton reasoned that if it fell and kept missing the Earth, it would go into orbit. So the idea of an orbit is simply continually falling under gravity but never hitting ground. The idea behind this is that gravity is a centripetal force, that you create an orbit by continually pulling the path toward the center. Here is an illustration of this.

Suppose you have a mass. We'll call this blue spot here the Earth. And then you have an orange spot here, which we'll consider something like the Moon. And it's moving perpendicular to the radius vector of Earth. And it moves along in its orbit, but at the same time, it's being pulled by gravity towards the Earth. So, while it moves up here it's pulled in this direction, and then its new position would be the sum of those two vectors, or here. It would then be traveling on a new perpendicular path, and it would be traveling in this direction, but again, pulled toward the center of Earth and moved to this direction.

That process would continually repeat, moving along its tangential path but yet being pulled back toward the center of the planet. And that repeats around the orbit, and you can see in this rather rough diagram that you end up working all the way around, creating an orbit. Now, this is a rough diagram because I've used a very long vector here, but gravity acts in a continuous manner, and so instead of having these 8 or 10 separate points, you would actually have a continuous orbit, and here we have a circle.

Now, we get a circle simply because I didn't want to go through the extra effort to try and make it into an ellipse. The circle is the easiest thing for me to create. But instead, if you took a more general diagram and more general motion, you would get an ellipse of this.

So, finally, this is what leads to Newton's law of gravity, and the law of gravity is stated thus: The force of gravity between two objects is proportional to the masses of the objects and inversely proportional to the square of the distance between them. I know that saying it in words doesn't quite get it out, so we use mathematics, and this is the equation: $F = GMm/r^2$.

F stands for the force of gravity, G for Newton's gravitational constant, M for the mass of one body, m for the mass of another body, and r for the distance between them. And so the force on the left-hand side is proportional to the mass of both objects, which are in the numerator on the right-hand side, and inversely proportional to the square of the distance between them. So as you have larger bodies, the force goes up. As you place them further away, the force goes down.

Newton's gravity work led him to discover his three laws of motion that he published in his book, the *Principia*. And those three laws of motion describe general forces, not just gravity. The important thing for the solar system is that Kepler's empirical laws can be derived from Newton's law of gravity. They are a direct result.

Gravity is an inverse square law, and if you do the mathematics to find out what is the general orbit shape for any inverse square force, you get an ellipse. The force is stronger when you are closer to the Sun and weaker when you're farther from the Sun. That leads to the observed variations in motion of the planets in their orbits around the Sun. Furthermore, larger orbits in an inverse square law have longer length, and they also have slower speeds, and that leads to the relationship between period and semi-major axis.

Newton's great achievement is that he created a physics foundation for the astronomical motions. As illustrated in this lecture, science is a search for order in the universe, and it's often a three-step process. First, you have to collect some data, then you're going to search for patterns in the data, and then you need to explain why such patterns occur.

In the heliocentric model, Kepler is the one who found the patterns, and Newton is the one who explained why they occur. The combination of these two efforts by these two great men confirmed the heliocentric model, and only with their work could we say the heliocentric model is the true model of the solar system.

And, actually, it goes a little bit beyond that because you're moving from the heliocentric solar system to what I think of as the gravitational solar system. You're going beyond just modeling things going on. You're creating a mathematical theory that describes the motions, and in it you're moving to a new way of thinking about the solar system, the mathematics underlying the physics that underlies astronomy. And here, for the first time, astronomy combines with physics, and we end up with a new science of astrophysics.

Well, that's it for today, and we'll see you next time. Take care.

Lecture Five
Planetary Predictions and Scientific Theory

Scope:

The true test of a scientific hypothesis reaches beyond explaining known observations into predicting the results of future observations. For planetary orbits, the Titius-Bode rule matches the orbits of the planets out to Uranus, and it appeared to predict the existence of the asteroid belt. However, the rule failed for Neptune and was relegated to an intriguing mathematical curiosity. Physical laws must pass stringent tests, and Newton's gravity was seriously questioned when Uranus defied all attempts to predict its orbit. Only when two astronomers postulated the existence of Neptune, and successfully predicted its position, were the observations reconciled. A similar test arose with minor deviations in the orbit of Mercury. This time, predictions of and searches for a new planet were fruitless. The theory of gravity had to be extended into Einstein's general relativity in order to explain the observations. With these examples, we see that the current knowledge of the solar system will continually be challenged by and must eventually adapt to new observations.

Outline

I. The value of a scientific hypothesis is evaluated by testing its predictions over time.
  A. An explanation is treated as a hypothesis until its predictions have been tested repeatedly by observations.
   1. In the public vernacular, the word "theory" has come to mean the same thing as "hypothesis."
   2. In scientific parlance, the word "theory" should be reserved for only those ideas that have proven true through rigorous testing over many years.
  B. The heliocentric model was not generally accepted until it could be used to make better predictions of planet positions than the geocentric model.
  C. Newton's ideas on gravity were likewise not immediately embraced as what today we call the fundamental law of physics.

- **D.** As astronomers test known ideas, they also search for evidence of new ideas.

II. A mathematical rule about the spacing of planet orbits turned out to be a numerical curiosity.
- **A.** The Titius-Bode rule approximated the distances of the known planets from the Sun.
  1. In 1766, Johann Titius noted the formula as a footnote in his book.
  2. In 1772, Johann Bode, a much more influential astronomer, included the rule in his book and began to argue for the existence of a missing planet.
  3. The rule states that the distance to Mercury is 0.4 AU, and the distances to the rest of the planets follow by adding a second term that doubles each time [Venus (0.4 + 0.3), Earth (0.4 + 0.6), Mars (0.4 + 1.2), unknown (0.4 + 2.4), Jupiter (0.4 + 4.8), Saturn (0.4 + 9.6)].
  4. The accuracy of the rule is 5% or better for the planets known at the time.
- **B.** The rule successfully predicted the distance of Uranus and the asteroid belt.
  1. In 1781, Uranus was discovered, and its distance agreed with the formula to 2%.
  2. The first asteroid (at the time called a planet), Ceres, was discovered in 1801, and fit the rule at a level of only 1%.
  3. Three other asteroids were soon discovered with similar orbital distances, making four planets between Mars and Jupiter.
  4. Bode's "law" became accepted as an empirical predictor of planet distances although no one had a good reason why it worked.
- **C.** The rule did not correctly predict the distances of Neptune or Pluto.
  1. In 1846, Neptune was discovered, but its distance was off from the prediction by 29%.
  2. Pluto, discovered in 1930, was off by 95%, though it fits with the Neptune prediction to better than 2%.
  3. The rule was shown to be generally not true, and it is today considered to be just an intriguing mathematical curiosity.

III. The law of gravity was severely tested by observations of Uranus but succeeded in predicting the existence of Neptune.
  A. For 60 years, calculations of the orbit of Uranus failed to predict the planet's position accurately.
    1. Within a few years of its discovery, searches of historical records found observations of Uranus dating back about 100 years.
    2. Astronomers calculated Uranus's orbit many times, but each time it strayed measurably from predicted positions in about a decade.
    3. Computing an orbit by hand is quite difficult, as the calculation uses the gravity of the Sun as well as perturbations due to the other planets (mostly Jupiter and Saturn).
    4. The problem of Uranus became a high profile challenge to astronomers, with some suspecting the influence of an unknown planet and others suspecting that Newton's gravity formula wasn't entirely correct.
    5. An important clue came in 1835, when Halley's Comet reached perihelion one day later than predicted.
  B. Two astronomers calculated the position of a planet that could create the deviations in Uranus's orbit.
    1. The problem is significantly harder than an orbit calculation, as it essentially requires determining two planetary orbits simultaneously.
    2. In England, John Couch Adams made the initial calculations, but strong personalities and terrible luck prevented any British astronomer from discovering Neptune.
    3. In France, Urbain Le Verrier's calculations were greeted with acclaim, yet no French observer would search for the predicted planet.
    4. Upon Le Verrier's request, German astronomers found Neptune in their first night of searching.
  C. Neptune's prediction created an international stir and was a triumph for the law of gravity.
    1. In France, celebratory pride was mixed with harsh accusations of England's trying to steal credit.

2. British astronomy went into an uproar as it uncovered both Adams's work and the reasons why the discovery was missed.
3. Astronomer Johann Encke called Neptune's prediction "the most outstanding conceivable proof of the validity of universal gravitation."
4. Mathematical prediction became a respectable foundation from which to guide astronomical observations.

IV. The orbit of Mercury deviates from Newton's gravity and requires Einstein's general relativity to explain it.
   A. The orientation of the orbit of Mercury rotates more than expected.
      1. The perihelion is the closest point to the Sun in a planet's orbit. The aphelion is the most distant point.
      2. The orientation of a planet's orbit slowly rotates, or precesses, due to the gravitational influence of the other planets.
      3. Observations showed that the perihelion of Mercury was precessing more than Newton's theory of gravity predicted.
      4. Modifications to the law of gravity were considered but were hard to believe after the success of predicting Neptune.
   B. Le Verrier predicted new planets to explain the discrepancy.
      1. The same mathematics that predicted Neptune was used to predict a planet or a collection of planets inside of Mercury's orbit.
      2. One astronomer claimed to have found an inter-Mercurial planet he called "Vulcan."
      3. Other astronomers searched on and off for 50 years but never found Vulcan or Le Verrier's planets.
   C. General relativity extended the law of gravity and accounted for Mercury's behavior.
      1. General relativity reinterprets gravity as mass producing dimples in the shape of space. Orbits are the natural motion through a curved space.
      2. General relativity extends gravity to handle cases of very large mass.

3. Mercury, as the closest planet to the Sun, is the only one to have its orbit changed noticeably by general relativity.

V. Predictions are a fundamental aspect of science that both test current ideas and lead to new knowledge.
   A. Sometimes predictions, such as the Titius-Bode rule, are just numerical coincidences.
   B. Sometimes predictions lead to new discoveries, as in the case of Neptune.
   C. Sometimes predictions indicate new physics, as when general relativity is required to explain Mercury's orbit.
   D. Science is a process of testing a hypothesis over many years until the community is comfortable calling it a scientific theory.

Suggested Readings:

Bennett, Donahue, Schneider, and Voit, *The Cosmic Perspective*, chaps. 3–4, 7.

Grosser, *The Discovery of Neptune*.

Hoyt, *Planets X and Pluto*.

Sagan, *The Demon-Haunted World*.

Weintraub, *Is Pluto a Planet?* chaps. 6–9.

Questions to Consider:

1. Can you think of a hypothesis that might explain the Titius-Bode rule? How would you test that hypothesis?
2. Adams predicted the existence of Neptune first, but Le Verrier published his prediction first. Who deserves credit as the discoverer?
3. Does the existence of general relativity mean that Newton's law of gravity is wrong?
4. If the testing of a hypothesis takes many years, do most scientists live to see their work elevated to the status of scientific theory?

Lecture Five—Transcript
Planetary Predictions and Scientific Theory

What's the difference between a "hypothesis" and a "theory?" In the public vernacular, not much. We say, "Here's my theory of what happened" or, "Oh, that's just a theory." But in science, the difference is huge. A hypothesis is an unproven assertion. It needs to make predictions, and those predictions need to be tested by observations, and if those observations match over repeated testing, then, and only then, does it get to become a theory. So the phrase "Oh, it's only a theory" would never be heard in science because to become a theory, you must be elevated up to become a theory.

The heliocentric model was not generally accepted until its predictions became better than the geocentric model. It was a hypothesis until it proved itself. Newton's law of gravity similarly took time for people to embrace that idea. We will show you one major example of testing of Newton's law of gravity in this lecture.

In science, we're always looking for new phenomena as a way of testing our old ideas and in discovering new ideas. We will find that established theories can be modified. We're going to discuss this in the context of predicting planets. There have been many predictions of planets throughout history. We're going to discuss three examples, each that have a different resolution.

For our first example, we're going to discuss the Titius-Bode rule. In 1766, Johann Titius noted a formula in a footnote in one of his books. It was a mathematical rule about the spacing of planet orbits, and it approximated the distances to the known planets pretty well. In 1772, Johann Bode, a much more influential astronomer, included the rule in his book, and it became widely known.

The Titius-Bode rule goes like this: We start with Mercury, and the Titius-Bode rule says that the distance to Mercury is 0.4 AU, remembering that an AU stands for an astronomical unit. It's the distance from the Sun to Earth. The distances to the other planets were found by adding a second term. For Venus, we add 0.3; for Earth, we add 0.6; for Mars, we add 1.2; and as you can see, the second term doubles each time.

And so if you take each of the planets as being labeled by the letter N, and you turn the Titius-Bode rule into a mathematical formula, the distance would be $0.4 + 0.3 \times 2^{(N-2)}$.

How well did it work? Well, the accuracy of the rule for the known planets at the time was 5% or better. And in this table, I give you the distance, as given by the Titius-Bode rule in the first column and the semi-major axis in the second column. And you can see that they roughly match. In the third column, I give you the percent error. And so for Mercury, the difference between prediction and what was observed was 3.4%; for Venus, 3.2%; Mars was the largest at around 5.3%; and then Saturn at 4.8%. Both Earth and Jupiter were predicted almost exactly.

You'll notice, however, that the $N = 5$ planet is missing. Is that an error in the hypothesis? Is that a prediction of the Titius-Bode rule that's wrong? Johann Bode interpreted this as a challenge, and he began to argue for the existence of a missing planet.

Fortunately for him, a planet was found, but it was not where Bode expected. Bode expected the $N = 5$ planet to be found between the orbits of Mars and Jupiter. Instead, in 1781, William Herschel discovers Uranus. This is outside the orbit of Saturn. This is not the $N = 5$ planet. Instead, it's the $N = 8$ planet. The distance for Uranus that was found when its orbit was calculated agreed with the Titius-Bode rule to better than 2%. This was an expected level of support for the rule and convinced people that the $N = 5$ planet must exist.

So several groups of astronomers went out and started searching for the $N = 5$ planet. They actually formed a group called the Celestial Police. They had to find the missing planet. In 1801, a planet was found, not by the Celestial Police but by someone doing a survey, and this planet was named Ceres. We know Ceres today as being the largest of the members of the asteroid belt, but when it was discovered in 1801, it was the missing planet. And it agreed, again, with the Titius-Bode rule at a 1% level. Amazing, now we have not only the $N = 5$ planet, which matches, we have a new $N = 8$ planet. This is really wonderful.

Things got a little bit more complicated when three more $N = 5$ planets were soon found. Pallas was discovered the next year; two years later, Juno; and a few years later, Vesta. Pallas agreed with the $N = 5$ position to 1%; Juno to 5%; and Vesta, well, it was off by 19%. But you could argue that they were all part of the same $N = 5$ planet, but they had broken into fragments. Since they were so small anyway, you could justify that the Titius-Bode rule still worked. You just had the one planet that was broken into four pieces.

Nevertheless, the Titius-Bode rule, as it was previously known, now became law. And this was the level of agreement at that time. We see Uranus, Ceres, Pallas, and Juno with their Titius-Bode rule distances, and the semi-major axis distances, and here are their percent errors. And yes, Vesta is big, but as we said, it was allowed because we had four planets in that area.

This was an empirical law, just like Kepler's laws were empirical. No one had any good reason why it worked, because if you look at Newton's law on gravity, it says nothing about the spacing of orbits. Many explanations were offered, but none of them were lasting because the rule was not a physical law.

It turns out that in 1846, Neptune was discovered, and its distance was off from the Titius-Bode rule by 29%. In 1930, Pluto was discovered, and its distance was off by 95%, so I can add them to this table. And you can see that although we fit nicely for the first six discoveries, the next two discoveries have very strong errors. And in retrospect, you can see that Pluto's distance of 39.5 AU sort of fits with the expected distance for Neptune of 38 AU, and so Pluto might sort of fit into the $N = 9$ slot—but to do that, you'd have to ignore Neptune, and Neptune's much too big a planet to ignore.

So the law went back to being a rule. We think of it today as a very curious mathematical coincidence, and I consider it more like Kepler's numerology, the stuff that he put into his book *Harmonices Mundi*. There was no physical justification needed when you're just doing numerology because these are wonderful coincidences. And so we've stopped searching for a physical reason as to why the Titius-Bode rule works, or shall we say, almost works.

Our second example of planetary prediction is the prediction of Neptune, and this derives from a problem with the orbit of Uranus. After Uranus was discovered in 1781, very soon thereafter, we found historical observations that dated back about 100 years. When you've got that long a baseline of observations, you can calculate a relatively accurate orbit.

The problem was that those who tried to fit the orbit had a really hard time fitting both the old observations and the new observations at the same time with one orbit. No matter what orbit they tried, Uranus strayed from the predictions. When they came up with a new orbit and predicted where Uranus would be, after a few years or a decade, Uranus would no longer be in that position on the sky.

What are we going to do about this? Does this indicate that there's a problem with Newton's gravity? This was over 100 years after Newton had come up with his theory of gravity, but people were still willing to question it. However, computing orbits by hand was really difficult, so you couldn't actually be totally sure. When you're doing an orbit by hand, you have to not just take into account the gravity of the Sun, but you also have to take into account the gravity of Jupiter and Saturn, and sometimes other planets, as they might slow or speed a planet in its orbit.

The problem with Uranus became a very high profile challenge, and in 1835 came a very important clue. Halley's Comet came back in 1835, and reached perihelion one day late. Now, you may think to yourself: Wait a minute, one day late? Halley's Comet has a 76-year orbit. What is one day in that? That's 1 part out of 27,000. But the idea was that since Halley's Comet has such an elongated orbit, it would sample the gravity of the outer solar system. And if it sampled the gravity of an unknown planet, that could perturb its orbit and cause it to come back to perihelion one day late.

The prediction and the discovery of Neptune is one of the greatest scientific soap operas I've ever heard. It's a wonderful story, so please, join me in it. It starts in England, where this man, John Couch Adams, made his predictions. He was a brilliant mathematics student at Cambridge, and he took up the challenge of Uranus after he graduated. Now, this is an extremely difficult challenge because you're not just computing one orbit—you're actually computing two orbits simultaneously. So you're getting the perturbations of the Sun, Jupiter, and Saturn on one orbit, and on another orbit, and then you have to compute the perturbations of each of those orbits on each other.

Incredible problem, but Adams tackled it, and in 1843 he came up with his first solution. That just fit the observations of the modern era. He continued to work on it and tried to include the observations from 100 years ago, and in 1845 he came up with his revised solution.

Now, it's very important here that he decided not to publish it. He didn't want to put it in some rinky-dink little journal or, you know, papers according to some small society—he wanted to go for the big time. And to do that, he had to go through the astronomer royal, George Airy. In September of that year, he tries to meet with Airy, but there was some really bad timing, and a little bit of miscommunication, and some strong personalities, and it turns out that they never meet.

He leaves his solution with Airy, but he feels rebuffed, and he feels a lot of bad feelings toward Airy. So now we turn to George Airy, and he picks up his turn in the story. As astronomer royal, he is basically one of the head honchos of English astronomy. He, however, has a bias against theoretical calculations, and he also has a bias against young researchers. John Adams was 26 when he produced this solution. Can a young man of that age really produce anything valuable? He didn't give it much credit, and he sent off a reply to Adams that, you know, asked some trivial questions about it.

But in October, he did show the solution to a man named William Dawes. Dawes thought it was incredible, and he sends it to the observer William Lassell. However, Lassell has a sprained ankle, and by the time he's recuperated, the maid destroys the letter. Lassell, most assuredly, could have seen Neptune at that time, based upon Adams's prediction, but he was unable to because he lost the coordinates to look at before he had recovered from his sprained ankle.

So England was stymied, due to some strong personalities and to a string of bad luck. Now, the story moves across the channel to France, and this man, Urbain-Jean-Joseph Le Verrier, takes up the challenge of Uranus. Originally, Le Verrier wanted to study chemistry, but he moved to astronomy when a job opened up. And in 1845, François Arago, the dean of French astronomers, asked Le Verrier to work on the problem of Uranus. And when the dean of the astronomy in your country asks you to, you do it. He dropped everything and immediately began work on the problem of Uranus.

In November of that year, he had his first paper ready, and what that paper did is it corrected the errors in Uranus's orbit. And by June of the next year, he had a second paper ready that had gone through the whole calculation and was able to predict the position of the planet. For the French astronomers, he presented his paper to them, and they called it an analytical triumph. They thought it was fantastic. This was wonderful, but none of them would go to the telescope and search for the planet. Le Verrier was totally frustrated.

So, now we go back to Airy in England. He receives a copy of Le Verrier's paper. He now knows that Adams's and Le Verrier's predictions agree to within 1° on the sky. Finally, George Airy asks the Cambridge Observatory to search for the new planet. The observatory is run by James Challis. James Challis has a strong personality, and he's had an antagonistic history with George Airy, and so there's a little bit of give and take between them.

Furthermore, Airy asks for a very specific search method. He wants Challis to sweep across the sky in these very long arcs. He doesn't ask him to go look where Adams predicted. He just says: Do a survey. It was a cumbersome method, but still Challis observes Neptune four times. However, again, more bad luck—he doesn't analyze the data. He analyzes data up to the observation, but doesn't actually go to the observations that contain Neptune.

Back to France: In August, Le Verrier produced a third paper, with an even more refined prediction of the planet. Again, no French astronomer would go out and look for the planet. He gave up on them. Finally, he had some correspondence with somebody at the Berlin Observatory, and he said: Hey, maybe they will look for it. He left his countrymen behind, and he sent a letter to Johann Galle at Berlin Observatory and asked him to search for it.

Galle got permission to use the observatory, and on the very first night they looked for it, he found it. The second night, they confirmed it, and they had the discovery. Neptune was found. This discovery led to an international controversy because Galle and Le Verrier announced their result. Airy, in England, has to save face. He has to start dancing around and saying: Wait, wait, wait—I knew of Adams's result before Le Verrier published even his first paper.

The French Academy of Science went ballistic. Why were they only hearing of Adams's claim of prediction after the discovery had already been made? François Arago has an especially strong quote:

> Mr. Adams has no right to figure in the history of the discovery of the planet Le Verrier either by a detailed citation nor by the slightest allusion. In the eyes of every impartial man, this discovery will remain one of the most magnificent triumphs of astronomical theory, one of the

glories of the Académie and one of our country's noblest titles to the gratitude and admiration of posterity.

Some fighting words from France. You also had some serious attacks in England. The Royal Astronomical Society met the next month, and they did their best to give Adams his due, but they could not avoid doing savage attacks on Airy and Challis. Why had they let the discovery go to the French? Then the press got hold of the story, and they inflated the story all out of proportion.

Matter of fact, here is a cartoon in one of the French papers, and you can see the great big giant dunce cap on Adams's head as he is discovering Neptune by looking in the calculations of Le Verrier. The press overreaction helped the scientists to actually calm down because the scientific result was absolutely amazing. This was an analytical triumph. One astronomer called it "the most outstanding conceivable proof of the validity of [the law of] universal gravitation."

Mathematics could not only explain the known solar system, it could predict entirely new features—and mathematical calculations became a respectable guide for basing your observations. As a small footnote to the story, in June of 1847, Adams and Le Verrier met for the first time, and rather than being antagonistic, as you might expect from the story, they quickly became friends and remained so for the rest of their lives.

The third example of planet prediction now deals with a problem in Mercury's orbit. The perihelion of an orbit that I showed in this diagram is the point in the orbit that is closest to the Sun. The opposite point is called the aphelion. That's the point where you're the furthest from the Sun. The perihelion of the orbit can precess due to the gravity of other planets. It can actually rotate around, as shown here. So the orientation of the orbit slowly shifts over time. This was known due to Newton's law of gravity.

The problem with Mercury was that the precession was too large. Newton's gravity predicted a shift of about 532 arc seconds per century, whereas the observed rate was around 574 arc seconds per century. The discrepancy was 43 arc seconds per century. Now, that's a tiny amount. Let's put this in perspective. A circle contains 360°. Each degree contains 60' (minutes) of arc. Each minute of arc contains 60" (seconds) of arc. So we're talking less than 1 arc minute per century is the discrepancy here.

And, after the success in predicting Neptune, it was very hard to doubt Newton's law of gravity. So, what are you going to do? You're seeing a change in the solar system. You're going to predict that there's a planet causing it. None other than Urbain Le Verrier himself predicted a planet interior to Mercury that could cause this shift in Mercury's orbit.

Actually, Le Verrier found that a swarm of planets, a swarm of about 20 planets interior to Mercury would be the best fit to causing the shift in Mercury's orbit without causing a shift in Venus's orbit. One observer actually claimed to have seen such a planet transit across the face of the Sun, and he called it the planet "Vulcan." So yes, the planet Vulcan not only exists in *Star Trek*, it also exists in your history book.

Unfortunately, this observation was never confirmed. Vulcan was never again seen in transit, and when there were solar eclipses, there were expeditions launched to find it, and it was never observed during solar eclipses. The only thing that was keeping the planet Vulcan alive was the fact that it had Le Verrier's name on the prediction. So after Le Verrier died in 1877, the idea faded away.

However, the problem with Mercury's orbit did not. Fast forward to 1916: Einstein's general theory of relativity is published in this year, and to give you a three-word summary of general relativity—mass warps space. This is the idea that the presence of mass changes the fabric of space-time. In this illustration, the blue grid is supposed to represent the fabric of space-time, and the yellow ball is supposed to represent the mass of the Sun. And the mass of the Sun warps space, so it's no longer flat. Gravity is no longer a force, but it's reinterpreted as geometry and as the curvature of space.

So, if you have a flat space, the normal motion through it is a straight line. But if you have a curved space, the normal motion through it would be an orbit. The analogy I like to give is think of a waterbed. What if you take a bowling ball and place it in the center of a waterbed? You're going to warp the surface of the waterbed. Without the bowling ball in the waterbed, you could roll a BB straight across it. It would travel in a straight line. But with the bowling ball in the center of the waterbed, as you roll the BB across it, it will go down into the depression made by the bowling ball and it will have a curved path through a curved space.

With this diagram, you can see the depression made in the fabric of space-time by the mass of the Sun, and the planets would be rolling around the interior of that depression in an orbit. So as one physicist put it: Mass tells space how to curve, and curved space tells mass how to move.

General relatively was able to explain the extra precession of Mercury's orbit because general relativity extends Newton gravity to handle realms where there are very large masses. And Mercury is the only planet close enough to the Sun to show these effects. So, the planet Vulcan finally disappeared in 1916, and it disappeared into the equations of general relativity.

In this lecture, we've gone through three different examples of planet predictions, with three different outcomes. When we looked at the Titius-Bode rule, it started to look like a law but ended up just looking like a numerical coincidence. The Neptune prediction started out by a challenge to Newton's law of gravity, and it ended up with a triumphant verification of Newton's law of gravity. The discrepancy in the orbit of Mercury, possibly indicating new planets, actually indicated new physics.

This is a challenge in science: that we don't know which hypotheses will pan out. Where do we put our efforts? There's always a number of really interesting ideas to pursue. Which ones will turn out to be coincidences, and which ones will lead you to new discoveries? We don't know those things.

So if we answer the question that we posed at the beginning, "What is the difference between hypothesis and theory?" the answer is simply years and years of confirming our observations—a lot of hard work.

That's it for today, and we'll see you next time.

Lecture Six
From Observational Evidence to Discoveries

Scope:

The teaching of science tends to celebrate the discoveries and end products of the scientific method, despite the fact that the observations and data gathering play just as important a role. This idea is especially true in astronomy, where the basic job of creating an accurate and detailed map of the heavens has been a continuing and evolving task for thousands of years. Careful and patient surveys of the sky not only uncover new knowledge but also serve as the basis of other observations in astronomy. Since the invention of the telescope, craftsmanship in their construction has conferred advantage in making discoveries. As telescopes improved and grew larger, they reached limitations due to size, weight, and Earth's atmosphere. Advances continued through the development of instruments attached to the telescopes that, for example, disperse light into its spectrum or capture it in a photograph. Observational improvements have continually enabled scientific advances. Each stage of research—instrumentation, observation, and interpretation—has developed into a specialized field for astronomers.

Outline

I. Our big ideas about how the universe works are based on a long chain of evidence and argument.
   A. The models of the solar system and physical laws are the end product of the scientific method.
   B. Gathering the data to develop and support those ideas is just as critical a component of research.
   C. As astronomy has developed as a science, specialization and collaboration throughout the discovery process have become more important.
   D. The romantic view of a lone astronomer on a mountaintop in a solo search for answers is far from accurate.

II. The data from pre-telescopic observations were mainly concerned with positional astronomy, or astrometry.

A. The motions of the Sun and stars define the day and the year.
   1. The line on the sky from north to south is called the meridian.
   2. The average time between the Sun crossing the meridian is the 24-hour solar day.
   3. Due to Earth's tilt and elliptical orbit, the Sun's motion across the sky varies over a year.
   4. The average time between a star's crossing of the meridian is 23 hours 56 minutes, which is called the sidereal day.
   5. The stars serve as a reference frame, and the positions of the Sun and planets can be related to which stars they are near or which constellation they are in.
B. Positions on the sky are defined in terms of the celestial sphere.
   1. Right ascension and declination on the sky are analogous to longitude and latitude on Earth.
   2. The zero point of the coordinates is the point in the sky where the Sun's path (the ecliptic) crosses the celestial equator on the spring equinox.
   3. Declination is an angular measure in degrees, minutes, and seconds of arc.
   4. Measuring the time of meridian crossing gave rise to right ascension being measured in hours, minutes, and seconds.
C. Star positions serve as the basis for making measurements on the sky.
   1. Constellations serve only as geographical markers on the celestial sphere.
   2. An important job in astronomy is to define a catalog of known star positions.
   3. The position of an object is then measured relative to a known star.
   4. Over centuries the positions of the stars changed, revealing the precession of Earth's rotation (the precession of the equinoxes).
D. In 2000, astronomy moved to using the International Celestial Reference System (ICRS).
   1. Positions in ICRS are measured relative to 212 distant radio sources.

2. The precession of the equinoxes will not affect ICRS coordinates.

III. Telescopes extended the ability to see objects that are small, faint, and far away.
- A. The two basic types of telescopes are refractors and reflectors.
    1. Refractors use lenses to collect and focus light.
    2. The largest refractors are limited by the length of the telescope and the weight and optical quality of the lens.
    3. Reflectors use mirrors to gather light.
    4. The largest reflectors are limited by the size and optical quality of the mirrors.
- B. Telescopes have several important characteristics that determine their usefulness for research observations.
    1. Magnification, or resolving power, is the ability to see fine details.
    2. The resolution of large ground-based research telescopes is limited by the scattering of Earth's atmosphere.
    3. The size of a telescope's primary lens or mirror determines its light-gathering power.
    4. The quality of the optical path of the telescope can affect both the above characteristics.
- C. Craftsmanship in telescope making is an important factor in scientific discovery.
    1. Access to the best and largest telescopes has repeatedly enabled astronomers to make important discoveries.
    2. Galileo's improvements to the spyglass brought about the birth of telescopic astronomy.
    3. William Herschel discovered Uranus using his own telescopes, the finest reflectors of his time.
    4. Telescope makers are an important branch of astronomers.

IV. While individual discoveries are trumpeted, one can argue that science benefits much more from general surveys of the sky.
- A. Creating surveys of the sky can both reveal and enable discoveries.
    1. Surveys establish a reference set of knowledge against which future observations can be compared.
    2. Herschel found Uranus while doing a survey of stars.

       3. Finding Uranus within historical observations allowed the problem with its orbit to be more readily identified.
       4. Neptune's discovery was aided greatly by a recent, accurate star map.
       5. The long and painstaking observations of surveys are important, if unglamorous, work.
   B. Survey data taken for one purpose are often used for different, unrelated research.
   C. A favored approach in astronomy is a targeted search.
       1. Some targeted searches use surveys to develop a list of candidates for further observation.
       2. Other targeted searches focus upon a particular region of the sky.
       3. Searches for solar system objects have often looked along the ecliptic for retrograde motion at opposition.

V. Over the years, the telescope has decreased in importance, while the instruments used with the telescope have grown in importance.
   A. As telescope design encountered limitations, improvements had to come elsewhere.
       1. Telescopes gather and focus light, while instruments process and analyze the collected light.
       2. Today's best observatories are defined by both the telescopes and the instruments attached to the telescopes.
   B. Sensitive detectors allow astronomers to analyze light in many ways that the human eye cannot.
       1. Prisms and other devices spread light into a spectrum for study by wavelength.
       2. Photography uses time exposures to record faint objects.
       3. Many specialized devices have been developed for astronomical observations.
       4. As an example, the blink comparator allows one to quickly switch between two observations, and it was used to discover Pluto.

VI. Astronomy has developed into a wide-ranging field with many specialties.
   A. Building and running an observatory requires a dedicated team.

B. The design and fabrication of sensitive instruments demand specialized talents.

C. Observational astronomers must be careful and patient to produce high-quality data.

D. Those who interpret observations and speculate about their physical basis are called theorists.

E. Excellence across all these aspects of astronomy is generally what leads to scientific advance.

Suggested Readings:

Bennett, Donahue, Schneider, and Voit, *The Cosmic Perspective*, chap. 6.

Hoyt, *Planets X and Pluto*.

Levy, *Clyde Tombaugh*.

Shu, *The Physical Universe*, chaps. 1–2.

Weintraub, *Is Pluto a Planet?* chap. 10.

Wilson, *Astronomy through the Ages*, chaps. 2, 6.

Questions to Consider:

1. Since you can only see at most half of the sky at one time, how would you measure the relative positions of stars on opposite sides? How would you work up a global solution for positions? Would using many telescopes and many observers across the globe introduce or suppress errors?

2. Some research telescopes are many decades old. Why would professional astronomers still be using such old equipment? How can it compete with current technology?

3. Given the speed of modern communications, astronomers can collaborate with others anywhere in the world. How does this help or hinder scientific progress?

Lecture Six—Transcript
From Observational Evidence to Discoveries

In teaching science, we tend to celebrate the discoveries: Galileo spying the moons of Jupiter, Cassini finding details in the rings of Saturn, or Herschel discovering Uranus. We have a habit of focusing on individual achievements, like Ptolemy's geocentric model, or Copernicus's heliocentric model, or Newton's law of gravity. And in doing this, we can lose sight of the huge swathe of general development in the field. The field is continually advancing, and these advances provide the foundation for the discoveries.

Newton famously once said, "If I have seen further ... it is because I have stood on the shoulders of giants." The milestones are the end products of a long chain of evidence and argument. In astronomy, this is particularly true. The results can depend on data taken decades or even centuries earlier.

Le Verrier and Adams made predictions of Neptune, but they could not have been made without Flamsteen's surveys over 100 years earlier. Further, as the field has developed, specialization within the field has also developed. Building telescopes, devising cameras and instruments, making observations, analyzing the data, interpreting results as physical theory: These are all subspecialties of astronomy that are almost fields unto themselves.

Collaboration is now required for discovery. Remember that Le Verrier and Adams could predict Neptune, but neither one of them had the skills to go to the telescope and actually do the search themselves. So the sort of romantic ideal we have in our society of the lone astronomer, alone on the mountaintop in a solo search for truth—well, that hasn't been true for hundreds of years.

And in this lecture, I want to focus on the data gathering, the wide range of observations that are necessary and how they enable breakthroughs in astronomy. To start, let's look at astronomy before telescopes. This was mostly concerned with positional astronomy, or what we call astrometry. Now, our astrometry is initially defined by the Sun and the stars.

If we look at the Sun—here's a wonderful time-lapse image of the Sun. And you can see it goes across the sky once per day. This defines our day. The meridian line, which is the line from north to south, is our local noon. And we define one day from when the Sun goes from the meridian to the next time it crosses the meridian.

The Sun's motion also varies over the year. This is a wonderful image, and it's called an analemma. It's a picture of the Sun taken at the same time every day over a year. Well, not every day. It looks like about 30 different days over the year. And you can see it forms this very nice shape—that it's highest in the sky in the summer, and it's lowest in the sky during the winter. And that fits because we have longer days in the summer and shorter days in the winter.

But the side-to-side shape shows you that the solar day is not constant. It's not always 24 hours. Sometimes it's a little bit more, sometimes it's a little bit less. Twenty-four hours is just the average solar day. When we look at the stars at night and we take a similar time lapse image, we can see that they also show circles on the sky, but the star motion is slightly different from the Sun motion.

The time between meridian crossings for the stars is 23 hours and 56 minutes, and we call this a sidereal day. It's the reason why there are different constellations of stars appearing in the evening at different times of the year. For example, Orion—that is a constellation we in the Northern Hemisphere know as a winter constellation. We speak of the Summer Triangle. If I see the great square of Pegasus up in the evening, I know it must be fall.

The stars can serve as a reference frame for the other motions. If you look at the solar versus the sidereal day, there's a four-minute difference. And over a year, it adds up to an extra rotation for the stars. That extra rotation is just due to the Earth's orbit around the Sun. Alternatively, you can think of the Sun moving through the stars once per year, and this is actually the derivation of how we get 360° in a circle.

The Babylonians noticed that the Sun moves about 1° per day. You can see that it's really 1/365 of a circle per day, but 360 was a very nice multiple, being a multiple of 2, 3, 4, 5, 6, and so on, so they used 360° in a circle.

When we look at the Sun over the course of a year, it moves across the sky against the background stars. And in this image, you can see that the path it takes goes against various constellations. These are the constellations we know as the constellations of the zodiac, and the path it traces out across the sky is called the ecliptic.

Now, the coordinates on the sky are based on the idea of a celestial sphere. On Earth, we think of lines of latitude and longitude. These are lines of latitude on Earth. The lines of latitude run around the globe, and they're measured from the equator. We have the northern latitudes and we have the southern latitudes. Conversely, these are the lines of longitude on Earth. The lines of longitude run pole to pole, and they are north-south lines, so they are meridian lines. The zero point is called the prime meridian, and it's arbitrarily defined to run through Greenwich, England.

The coordinates on Earth are defined like this. We have the line from the North Pole to the South Pole going through the Greenwich meridian as our zero point of longitude, and our longitude is measured at an angle with respect to that. And then with respect to the equator, we measure our latitude upward.

The same coordinate system is used for the celestial sphere, but instead, latitude becomes a measure called declination, and longitude becomes a measure called right ascension. We have, on the celestial sphere, a northern celestial pole and a southern celestial pole. The zero point of declination is still the celestial equator, and things are measured north-south from the celestial equator—but the zero point of right ascension is where the equator and ecliptic intersect, and that is the vernal equinox, when the Sun passes through the equator going northward in the spring. So, the right ascension is measured with respect to the vernal equinox.

Now, the declination is measured in degrees, minutes, and seconds, but right ascension is measured in hours, minutes, and seconds. And to understand that, you have to understand a little bit of observing. This image here is of the famous observer Tycho Brahe. This is his great mural quadrant, and you can see that what you would do with it is that you would align it along your meridian, the north-south line, and you can see the angular measures along it, and you would use it to measure the angle above the horizon of a star or planet as it crosses the meridian.

This was one of the greatest astronomical instruments before the invention of the telescope, and really, it's basically just a great big protractor. You would also measure the time of crossing of the meridian. And it was easier to measure the time than any angle with respect to the prime meridian, so that time became the convenient unit for measuring right ascension. Today, however, we find it to be an annoyance, and we generally take our right ascension measures and convert them to degrees.

The star positions are the basis for our measurements on the sky, and really, although you hear a lot about constellations, those really don't matter much to astronomers. They're just geographical markers on the sky. If you're using the human eye to look at the sky, there are familiar reference patterns, but they're not much help in how we do our orientation for professional astronomy.

So, to use the stars as a reference system, it's an important job to define a catalog of star positions. You need to have some known stars against which you can measure other things. The position of all other stars would be measured relative to those known stars. Astrometry is when you're able to obtain a global solution of star positions. You have your set of known star positions. You measure those star positions relative to one another, and in fact your known star positions end up being measured relative to one another.

You can only measure relative star positions. You can't see the entire sky at once, and so the absolute solution across the whole sky must be calculated. You need to use large surveys and large catalogs. These are required just to define the coordinate system on the sky.

One thing that we found out when we started creating these catalogs was that the positions of the stars were changing. Now, you might think this is natural. Maybe the stars are actually just moving. Well, yes, that's true. They are moving, but what really was moving was the vernal equinox. The zero point of our coordinate system was changing, and this was due to the precession of Earth's rotation.

Now, if I spin this top here, you can see that its axis is wobbling. It's spinning around, but the axis which it's spinning around is wobbling around. Earth is also spinning on its axis, and its axis is also wobbling. This is the precession of Earth's rotation axis, and it causes a change in where the celestial north pole points on the sky.

This diagram shows the circle around which the North Pole and the celestial north pole moves throughout the sky. Currently, it's right near Polaris. Polaris is called the polestar. But about 4000 years ago, it was near the star Thuban; and in about 5000 years from now, it'll be near the star Alderamin; and in almost 10,000 years, it'll be near the star Vega. Now, when the pole is over near Vega, it's going to look kind of stupid to have a star named Polaris that's going to be many degrees away from the pole.

But that's what Earth's axis does. Over 26,000 years, it rotates around the sky. This precession means that the coordinates we give a star needs to be tagged with a date because the coordinates are relative to a defined equinox, and the equinox is changing. For my work, I've had to deal with B1950 coordinates or J2000 coordinates, and these are the coordinates relative to the 1950 equinox or the 2000 equinox.

Now, I'm about to tell you something that says sort of what I just said was a lie, because recently we switched out of that coordinate system. All right? You need to know the idea of precession of the equinox because it's true and it happens, but our coordinate system has recently become rationalized. We switched to something called ICRS, the International Celestial Reference System. It was defined to match the J2000 equinox, but it's going to retain fixed axes in space, so even though the North Pole will not point to Polaris in a few thousand years, the coordinates of Polaris will not change because the orientation of ICRS is defined by some very distant radio sources. There are about 212 defining sources, and they're mainly quasars that are way, way, way across the universe and won't change their position relative to us. We're no longer truly using the Sun or stars to define our coordinate system, even though historically that's the way they have been defined.

Well, that completes a summary of the pre-telescopic concepts. Now, let's turn to how telescopes extend our capabilities. Telescopes help us see objects that are small, faint, and far away. There are two basic types of telescopes: refractors and reflectors.

The refractors use lenses to collect and focus light. The classical idea of a telescope would be a spyglass. You have a spyglass, it has a lens at one end and a lens at the other end, and you're used to seeing people look through it. Well, that's the sort of telescope that Galileo used, only his was about 40 inches long. It had a lens at one end to collect the light and a lens at the other end to focus the light to look through it.

The largest refractors are limited by various factors, one being the length of the telescope. It was hard for Galileo. You could not hold a 40-inch telescope steady. You of course had to put it on a tripod. If you were trying to get even larger ones, you have to have very large mouths. They are also limited by the size of the lens, how big a piece of glass you could create, the optical quality of that lens—can you keep a large piece of glass with a very clear quality through it?—and the weight of the glass. Once you get a piece of glass this big, it starts to sag under its own weight.

This image here is an image of the Yerkes 40-inch telescope. Now, it's not 40 inches long. It's actually the lens at the end of it is 40 inches across, and it's 62 feet long. It is the largest refracting telescope in the world. The lens is 40 inches across and weighs 1000 pounds. To give you a sense of scale, here's a picture of some famous astronomers at the Yerkes telescope. And if you can look at it, you can definitely see that a lone astronomer could not operate this telescope on his own. You definitely need somebody else to help you run this telescope.

The second type of telescope, reflectors, used mirrors to gather light. This idea actually originated with Isaac Newton. He was frustrated with the quality of images you got from a refracting telescope and decided to try using mirrors to gather the light. This is a basic diagram of a reflecting telescope, and the light comes in the barrel down here, reflects off the mirror, comes back up to the secondary mirror here, and goes out through the eyepiece. This is the primary mirror at the bottom, and the secondary mirror up here in the tube.

This allows you to collect light more efficiently because mirrors bounce the light, and they also do not suffer from chromatic aberration, which is one of the main problems that Newton was trying to solve, in that the different wavelengths of light are refracted slightly differently in a lens whereas they are handled equally by a reflecting telescope.

The other interesting fact about these telescopes is you'll notice this secondary mirror, just hanging out here in the middle of the tube. Something has to hold up that secondary mirror, and usually you have some struts holding the secondary mirror. Those struts are the source of all these diffraction spikes on stars when you see them in telescope images. All of these crosses that you see are really diffraction around the struts holding up the secondary mirror of the telescope.

The largest reflectors are limited basically by the size of the mirror. You can build a cabinet to hold the telescope, almost as large as you want depending upon the machineries you've got, but you're really limited by the size of the mirror you can make, and of course by the optical quality of the mirror. But we're really good at making very good mirrors.

This image here is the Lord Ross 72-inch telescope. It's nicknamed the Leviathan. It's 58 feet long and has a 6-foot mirror. The mirror itself weighs 3.5 tons. It was quite unwieldy, but they devised a novel way of mounting it so that they could make use of it. It was not able to point anywhere across the sky. Instead, these walls were constructed around it, and it mostly pointed north and south. And you can get stars about 1 hour right ascension before meridian and 1 hour right ascension after the meridian.

The mirrors on these telescopes are quite large, and this is an image of the Palomar 200-inch mirror, one of the largest mirrors ever made. And they were taking this mirror off from the telescope in order to polish it, and you can't really get a sense of scale from here, but while they were polishing it, they took a picture of it. You can see just how large this mirror is. Every few years, they have to take it down and repolish it, so this is the 200-inch mirror at Palomar.

The telescope characteristics determine the usefulness of the telescope for scientific research, and there are several that are the most important. The first is the magnification or the resolving power. This is your ability to see fine details, and it's measured in the smallest angle that can be resolved. For comparison, the human eye can resolve details about as fine as 1 arc minute, remembering that there are 360° in a circle, 60 arc minutes in a degree, 60 arc seconds in an arc minute. So the human eye can see about 1 arc minute.

And telescopes are generally measured in arc seconds. For ground-based telescopes, it really doesn't pay to try and build a telescope that can get better resolution than that because you would be limited by the sky. When we have twinkle, twinkle, little star, it makes the stars beautiful, but it drives astronomers crazy because that scattering of light blurs the light from distant objects, and we can't really get from the ground much better than about 1 arc second, unless we use some special techniques.

The second characteristic of a telescope is its light-gathering power, and this is mainly just determined by the size of its primary lens or primary mirror. The more light you gather, the fainter objects you can see, and these fainter objects are generally more distant objects.

The third characteristic would be the quality of the optical path. You see, every time light refracts or reflects, some of the light is scattered, and you need high quality optics to both focus the light cleanly and to retain as much of the light as possible. You're building a giant telescope in order to collect those valuable photons from these very distant objects, and you want to make sure the optical path is as clean as you can get it so you retain as much light as possible for your research.

The craftsmanship in building telescopes has become a very important factor in scientific discovery. It has been shown throughout history that access to the best and largest telescopes has enabled new discoveries. Galileo improved the idea of a spyglass and was immediately able to see things that no one else had ever seen. William Herschel built the finest reflectors of his time and allowed an amateur astronomer—he was not a professional—to discover Uranus. He could see stars that some of the professionals could not see.

The telescope makers have become an important branch of astronomers and contribute greatly to the research discoveries. It's also important how you use the telescope. While individual discoveries are trumpeted, I would argue that science benefits more from general surveys. Surveys establish a reference set of knowledge against which future observations can be compared. And you don't know the benefits of it while you're doing the survey.

Herschel found Uranus while doing a survey. And a search of previous surveys finds more observations of Uranus, and then we identify the problem with Uranus's orbit. And that leads to the discovery of Neptune. What's further, Neptune's discovery was aided by a recent accurate star map that allowed them to find: Hey, this star's not on the map!

So while the survey work may be unglamorous—it's long, it's painstaking observations—and you may not see immediate benefits, the surveys are very important because the benefits extend throughout history. And often, surveys that are done for one purpose end up being used for different, unrelated research.

Now, sorting through an entire survey can be somewhat like looking for a needle in a haystack. Another standard approach that we use is the targeted search. We'll use a survey to develop a list of candidates and then use follow-up observations on those candidates. In fact, many surveys these days are not really for the general good of astronomy. They're primarily to find candidates for follow-up study.

We can also limit the data that we're looking through by searching one small area of the sky. If we look through that one small area thoroughly, and we do all our research there, we can get detailed results. But of course, that means we then have to assume that that area of the sky is a representative sample of the sky, and we have to then extrapolate the results from one region to the whole region of the sky.

For solar system objects, one search technique that works well is to search the area of the sky that is opposite the Sun. If the Sun is over here, we're going to examine this part over here, and what we're going to look for is retrograde motion because if there's a solar system object at opposition, outside the Earth, it will be going through retrograde motion. By looking in that over the course of a year, we sweep through the ecliptic.

Finally, and this is probably the most surprising point that I will make today, telescopes really don't matter that much any more. I know I'm speaking crazy here now. What really matters when we go to a telescope is the instruments that we put on the telescope that process and analyze the light. The telescope's primary job is just to gather a lot of light. We've got really good telescopes for doing that, and now what differentiates different observatories is the instruments that we put on them that enable the deep research.

So, let's go through a few examples of the instruments. The simplest instrument would be a prism, and here's a picture of a prism, where the light comes in one side and it's broken into its constituent colors on the other side. This allows us to look at the light of a star or other object and examine each color separately. Prisms are really just a basic idea of how we disperse light into its component colors.

There are also diffraction gradings and many other methods for looking at the dispersion of light. When you disperse the light far enough, you can actually see lines in the spectrum. In this diagram, this drawing, really, the light from a star comes through a prism and then is dispersed out. And you can see within the component colors, there are lines across it. These are called spectral lines, and these are the absorption of specific elements. As electrons change levels within the elements, they absorb specific wavelengths of light, and that reveals the chemical composition.

If we look at stars, this is a group of stellar spectra, and each of the spectra across here for the various stars shows you various spectra lines. Some of the spectra lines are the same—for example, this hydrogen delta line appears in the top three spectra. Some of them are different. What's fascinating is that we can actually study the makeup of stars by their light. We can study what's in a star, even though we can't bring it into the laboratory and dissect it and take it apart. What's also interesting is that we can discover new elements. The element helium was discovered in the spectrum of the Sun—*helios* meaning, of course, "Sun."

Another instrument that we use all the time in astronomy is photography. Photography was a major advance for astronomy because it allowed us to accumulate light. The human eye, when you look at something, only accumulates light for about 1/30 of a second. You can tell this is the secret of why movies work because movies are just a sequence of images, playing through a projector, and they're played at about 24 frames a second. When you play the images through fast enough, they look continuous to the human eye.

So, when we look at a distant object through a telescope with the human eye, we're only seeing 1/30 of a second of that light. If we want to see fainter details, we need to record that light. So, as an example, let's take a look at this image sequence. This is an image of an object, a 1-minute exposure. Taking a 1-minute exposure, it looks like this, and it's sort of a blobby thing. If we expose instead for 5 minutes, we see a little bit more detail. For 30 minutes, we see a lot more detail; and for 45 minutes, we can recognize it as the Andromeda Galaxy.

You need to use short exposures to expose the bright regions, and long exposures to see the faint regions. The telescope expanded what the human eye could see in terms of the size domain, how much light we could gather, and photography expands what we can see in the time domain, how long we can gather that light.

Many other specialized devices have been employed for astronomy. One example that I will go through is the blink comparator, and it allows you to quickly switch between two observations. This is an image of Clyde Tombaugh using the Zeiss blink comparator at the Lowell Observatory. The lever that he's holding allows him to switch between one image to another image. And what Tombaugh did is he would take images of the same part of the sky on separate days and then pull them into the blink comparator and look for things that moved. This was in the great search to find Pluto.

These are actually the images that Tombaugh blinked in order to find Pluto. And this was one image he took on January 23, 1930, and this is an image he took on January 29, 1930. If I go back and forth between them and you look really closely, you can see things moving there. Now, it's kind of hard to see them because at this resolution they're very small, but I put red boxes around about 10 objects where I see things moving between the two images.

If I blink back and forth, we can say, "Okay, the 23rd and the 30th. The 23rd and the 30th. The 23rd and the 30th." And you can see several of those objects moving. Which one is Pluto? Well, Pluto is this one here and that one there. How did he find it? It's the one undergoing retrograde motion at opposition.

So, astronomy has developed many different specialties. Building and running an observatory requires a dedicated team of people. The design and fabrication of sensitive instruments demand specialized talent. Observational astronomers must be careful and patient—and those who interpret those observations and speculate about their physical basis, those are the ones we call theorists. By its nature, astronomy has become a collaborative enterprise, and it is excellence across all these aspects of astronomy that generally leads to scientific advance.

That's it for today. We'll see you next time.

Lecture Seven
The Space-Age Solar System

Scope:

At the dawn of space exploration, astronomers had a well-defined overview of the solar system. The Sun has a collection of planets, moons, asteroids, and comets orbiting around it. These are spread across a scale so vast that it can be only partially breached by telescopes and photographic plates. New dimensions were revealed when astronomers developed telescopes in other wavelengths, such as radio waves and infrared light. The advent of computers and digital technology created new and faster ways to observe, analyze, and simulate the heavens. The most amazing details are those returned by an armada of satellite missions sent to fly by, orbit, and land on these diverse worlds. In just a few decades, the richness of our knowledge expanded exponentially. The golden age of solar system exploration had arrived.

Outline

I. The space age transformed our view of the solar system.
   A. The space age began on October 4, 1957, when the first satellite, Sputnik 1, was launched into orbit for 22 days.
   B. NASA was founded one year later.
   C. The ability to explore the solar system led to a knowledge explosion.
   D. At the start of the space age, our model of the solar system was well established.
      1. The contents included the Sun, planets, their moons, asteroids, and comets.
      2. The orbits are a mixture of almost circular to highly elliptical shapes confined mainly to the plane of the ecliptic.
      3. Planets complete their orbits in timescales from months to centuries.

II. The scale of the solar system is vast, almost too vast to be comprehensible.

- **A.** It helps to consider a scale model of the solar system in which the Sun is the size of a basketball: 9 inches (23 cm) in diameter.
- **B.** The sizes of the planets are much smaller than most expect.
    1. Jupiter is a little less than 1 inch in diameter (2.4 cm), and Saturn about 3/4 of an inch (2 cm).
    2. Uranus and Neptune are both about 1/3 of an inch (0.8 cm).
    3. Earth is less than 1/10 of an inch—only 2 millimeters in size.
    4. Venus is also 2 millimeters, with Mars just over 1 millimeter in diameter.
    5. The rest are submillimeter size, including Mercury (0.8 mm) and Pluto (0.4 mm).
    6. From this model it is easy to see that the Sun contains 99.86% of all the mass in the solar system.
- **C.** The orbital distances in this scale model strain the imagination.
    1. Putting our basketball Sun on the goal line of a football field, Mercury (0.8 mm) is at the 21-yard line, Venus (2 mm) is at the 39-yard line, Earth is past midfield on the opponent's 46-yard line, and Mars is on the opponent's 20-yard line.
    2. Our 1-inch Jupiter is more than 2.5 football fields (260 m) away, and 0.75-inch Saturn is about 5 football fields (470 m) distant.
    3. Uranus is a bit less than 1 kilometer out (0.94 km), and Neptune a bit less than 1 mile (1.5 km).
    4. Pluto, less than 1/2 a millimeter in diameter, is 2 kilometers from the basketball-sized Sun.
    5. Examining a scale model emphasizes that the solar system is mostly empty space.

III. The space age brought about a revolution in our observation of and understanding of the solar system.
- **A.** Advances in technology enabled astronomers to build telescopes in other wavelengths of light.
    1. Visible light is but a small part of the electromagnetic spectrum.
    2. Infrared and ultraviolet observations of the planets reveal details unseen in visible light.

3. The Sun takes on an entirely new visage in the high energy radiation of ultraviolet light and X-rays.
4. Studying all wavelengths of light gives astronomers a more complete picture of the physical processes taking place.

B. The digital technology of computers revolutionized astronomical research.
1. Conventional photographs do not have a simple relationship between the amount of light received and the brightness of the image.
2. Digital images provide a quantified numerical measure of the light received.
3. Computer processing of digital images can bring out details that are impossible to extract from a photograph.
4. Computer analysis of data and physics simulations opened up new fields of research.

C. Computer-controlled telescopes can be larger, more accurate, and easier to operate.
1. Professional telescopes are now run by observatory staff, not astronomers, and can sometimes be accessed over the Internet.
2. Computers enable multiple mirror telescopes by adjusting the mirror segments many times per second.
3. Computers allow arrays of smaller telescopes to combine their observations and function like one larger telescope.

D. Telescopes in orbit bypass the limitations of Earth's atmosphere.
1. Earth's atmosphere blocks most wavelengths and can distort the light that does pass through.
2. Satellite telescopes both improve observations and enable observations not possible from the ground.
3. NASA's "Great Observatories" are Hubble in visible light, Compton in gamma rays, Spitzer in infrared, and Chandra in X-rays.
4. A fleet of more specialized missions has covered the wavelengths from radio waves to gamma rays.

E. Manned and satellite missions across the solar system uncover details that enrich and transform our scientific views.

1. The Apollo missions to the Moon made space a visceral reality.
2. The Voyager missions made a grand tour of the outer planets.
3. A fleet of Mars missions has orbited, landed on, and explored the red planet.
4. The numerous satellite missions have transformed those tiny spots of light, so very far away, into known worlds and familiar sights within our neighborhood in space.

IV. The decades of the space age have brought us new ways to observe, analyze, and explore the solar system.

    A. Historians generally cite the Copernican revolution as the biggest change in perspective on the solar system.

    B. I disagree: I feel that the space-age transformation from observers to explorers has changed our perspective more than anything else in history.

Suggested Readings:

Bennett, Donahue, Schneider, and Voit, *The Cosmic Perspective*, chaps. 6–7.

Benson, *Beyond: Visions of Interplanetary Probes*.

Hey, *Solar System*.

McFadden, Weissman, and Johnson, *Encyclopedia of the Solar System*, chaps. 35–42, 46.

Moore and Arnold, *Space: The First 50 Years*.

Wilson, *Astronomy through the Ages*, chap. 9.

Questions to Consider:

1. The nearest star is about 250,000 times farther from the Sun than Earth is. Given the Sun as a basketball-scale model, how far away is that second basketball?
2. Different wavelengths of light show different objects and features of objects on the sky. How would you cross-correlate positions in images that show little similarity?
3. How should we spend our limited science satellite budget: on a few expensive general-purpose space telescopes or on a larger number of cheaper special-purpose missions?

Lecture Seven—Transcript
The Space-Age Solar System

October 4, 1957: This was the beep heard round the world. The Soviet Union had launched Sputnik 1, an 84-pound satellite, just a 23-inch sphere, but it orbited Earth every 98 minutes for 22 days. The beeping could be heard by amateur radio operators, and it spread the immediate message that the space age had begun.

For astronomy, no other event in history changed our view of the solar system like the ability to explore it. It is not a coincidence that NASA was founded almost a year to the day after Sputnik 1 launched. Considering the magnitude of the knowledge explosion that happened during the space age, it's helpful to explore what we knew at the dawn of the space age and take a look at the technological advances that would enable a plethora of new discoveries.

By the mid-1900s, our model of the solar system was well established. There was the Sun, the nine planets (remember, Pluto was still a planet back then), their moons, some asteroids, and some comets. And most folks' idea of the solar system was one of the orbits of the planets around the Sun. I have an interactive 3-D program here that I can use to illustrate this idea.

We start by looking at the orbits of the inner planets, the rocky planets: Mercury, Venus, Earth, and Mars. And you can see that they are ellipses, or nearly circular ellipses, around the Sun. The rocky planets are in close, but if I back away, you can begin to see the outer planets and Pluto: Jupiter, Saturn, Uranus, and Neptune.

And you can see that Pluto's orbit looks like it crosses Neptune's orbit, but that's really just a projection effect. It's a highly inclined orbit, and if I put things edge on, you can see that the four giant planets are lying mostly in a plane, while Pluto lies distinctly out of that plane. If I wiggle it back and forth just a little, you can see that Pluto's orbit doesn't really cross Neptune's orbit. It is totally distinct from Neptune's orbit.

Let's go back into the inner planets, and you can see that they too lie in a plane, so all of the planets basically lie in the same plane as Earth's orbit. If we go back to the face-on view and turn on time, you can see that Mercury orbits much faster than Venus, which orbits faster than Earth, which orbits faster than Mars. This is the Keplerian motion that comes from the universal law of gravitation.

As we go out, you can see the same thing is true for Jupiter. Let me turn on time a little faster here. Jupiter moves faster than Saturn, moves faster than Uranus, moves faster than Neptune—and slowest of all is Pluto. The planets orbit in timescales of a few months for Mercury, up to a few centuries for Pluto. Now, what that interactive couldn't show you was the size scale of these objects, and the size scale of the solar system is really vast. It's almost too vast to be comprehensible. So I like to shrink the solar system down to a manageable timescale in order to understand it.

My favorite comparison is, what if the Sun were the size of a basketball? A standard basketball is 9 inches across. Well then, what would be the sizes of the planets? Jupiter would be a sphere about 1 inch in diameter. Saturn would be about 3/4 of an inch, or 2 centimeters. Uranus and Neptune are both about 1/3 of an inch across, or a little less than a centimeter.

Earth is very small, and it's only 2 millimeters in size. Actually, I had to put it inside some pieces of tape, otherwise I'd lose it because it's so small. That is the true size of our planet compared to a basketball Sun. It's really only about the size of these little nibs, these tacky nibs to give you a grip on the basketball.

Venus is about the same size as Earth, again, 2 millimeters in size. Once you get smaller than that, I couldn't find any spheres to fit things, so for Mars, I found a large paperclip, and if you hold it up edge on, that's about the size of Mars. Mars is only 1 millimeter in diameter in this scale model. Mercury, well, that's about the size of a small paperclip, a normal paperclip. It's less than a millimeter in size.

And finally, Pluto. Pluto is so small that the only thing I could find was a piece of pencil lead, 1/2 a millimeter in diameter. That's really small. So, that's it. That's the scale of the solar system—a basketball Sun, we've got a 1-inch Jupiter, all the way down to a 0.5-millimeter-size Pluto. From this, you can easily see that the Sun contains 99.86% of all the mass in the solar system. Basically, you can forget all this stuff. This is just flotsam and jetsam. This is really the whole solar system.

Now, the natural question asked here is, "How far away would I have to put these planets in order for them to be at the right distances?" In order to get enough space, I would have to use a football field. Here is a diagram of this basketball-Sun-based solar system laid out on a football field. If the Sun is on the goal line, then our millimeter-sized Mercury is out at the 21-yard line. Venus is out at the 39-yard line. Earth is across midfield at the opponent's 46-yard line, and Mars is all the way 80 yards away, on the opponent's 20-yard line.

These tiny, millimeter-sized objects are spread out across the length of a football field. When you go out to the outer solar system, then you're going much further away, and our 1-inch Jupiter is more than 2.5 football fields away. Three-quarter-inch Saturn, 5 football fields away. Uranus is a bit less than a kilometer, and Neptune is a little less than a mile. Pluto, which is only half a millimeter in diameter, is over 2 kilometers away from this basketball-sized Sun.

So, you have to try to imagine that. For 4 kilometers across, there's just one basketball, a couple big-sized marbles, a couple small-sized marbles, some really tiny BBs, and then a bunch of really small things, and even smaller than that, and then there are the asteroids and comets. They're just flotsam and jetsam. They don't even add up to a good small moon. The solar system is really mostly empty space.

It's kind of amazing that we've actually been able to traverse it. In the space age, we were able to extend across the solar system. How did we do that? Well, with a lot of technology. Technology has really brought about a revolution in our ability to observe and understand the solar system.

Some of this technology allowed us to build telescopes in other wavelengths of light. You see, we see in visible light, but visible light is just a tiny fraction of the full electromagnetic spectrum. Visible light is depicted here in the middle of the electromagnetic spectrum, and it has a small range in the middle. But extending onto very long wavelengths, we have radio light, microwave light, and infrared light—and going to shorter wavelengths, we have ultraviolet light, X-ray light, and gamma ray light.

Light is really just the same phenomenon, whether it's gamma ray light or radio light—it just has a different wavelength, and that's what this wave is supposed to depict here in the middle of the diagram. The wavelengths for gamma rays go all the way down to 10^{-13} of a centimeter, whereas wavelengths for radio light go on up to a kilometer or larger. So it spans a huge range, and visible light is just about a factor of 2, a very small region right in here, in the center of that.

When we use telescopes in these other wavelengths, we see a lot more of the universe. For example, when we look at Venus in visible light, it's rather bland. It's kind of boring because it's entirely cloud covered, and it really just reflects most of its light back to space in the visible. But if you use ultraviolet light, you're able to see through those clouds and actually start to see structure in the atmosphere of Venus. This image here I know has to be in ultraviolet light because I can see the cloud structure in the atmosphere of Venus.

If we look at the Sun in visible light, again, like Venus, it looks a little bit bland. It's basically a nice smooth sphere, and maybe it has a few dots on it, the sunspots on it. These don't look terribly impressive. However, when you look at it in X-ray light, the Sun comes alive. Those sunspots are areas of fierce magnetic storms and are emitting huge amounts of energy in ultraviolet and X-ray light.

By studying all the different wavelengths of light, it gives astronomers a more complete picture of the physical processes taking place in our universe. Another major technology that has helped astronomy is digital technology—computers. When we used photographs for astronomy, they were an analog medium. The emulsions of a photograph respond in a complex way to the amount of light that is absorbed.

When a photograph absorbs twice as many photons of light, it's not necessarily going to be twice as bright. And as astronomers, we need to know how much light has been absorbed from an object in order to understand it. So we had to measure what we call the response curve of a photograph's emulsion.

When we moved to digital images, we bypassed that, and we were able to get a quantified, numerical measure of the light received. We use something called CCDs, which stands for "charge-coupled devices." These are the same things that are in your digital camera. You might even have a digital camera on your phone these days. Well, astronomers have been using CCDs for about 20 years, and they were actually an important secondary market for the makers of CCDs, so astronomers, in a way, helped bring about the camera that you may have in your phone today.

And once our data has been recorded digitally, we have it available for computer processing, and that can bring out details that are impossible to extract from a photograph. Now, for example, this is an image of a nebula, and it is done at a low stretch so that we can see the details in the very bright regions in the center of it. If, instead, I computer processed it to bring out the faint details, well, then the central region would be overblown, but I could see the details in the outer region. You'll see how much nebula there really is.

We can then work with it to try and come up with a balanced stretch digitally—and come up with an image like this. We can also take multiple images of it and digitally combine them together so that we can get a color composite of various wavelengths, looking at it in different filters of red, blue, and green—also filters that respond to different hydrogen atoms or calcium atoms or helium atoms. You can get a wonderful color picture like this.

Using computers also opens up ideas for using simulations. We have supercomputers that can process data very quickly, and we can take ideas that we could never observe in nature and simulate them on a computer. For example, there are two images of a galaxy collision sequence. In the top image, you see the two galaxies before they've collided. They are trapped in each other's gravity, and they're about to smash through one another. The bottom image shows the computer simulation after the two galaxies have collided. This simulation was run on a supercomputer about a decade ago and took about a week's worth of calculation time. In nature, the process would have taken about a billion years, but we can simulate it on the computer and study it.

Now, I know this is, of course, in the solar system, and these galaxies aren't part of the solar system. But trust me when I say we use computer simulation all the time in solar system objects as well, specifically for calculating orbits and observing the stability of the orbits within the solar system.

Computers are also very useful in working with our telescopes. Computer-controlled telescopes are easier to operate. They can be larger, and they can be more accurate. Matter of fact, professional astronomers never move a telescope any more. For one, there are telescope operators on the mountain, and number two, even the telescope operators don't move those big telescopes any more because they're all controlled by computers. Sometimes you don't even have to go to the mountain. Sometimes you can actually control the computer over the Internet, and you don't even have to leave your research institution.

As an example of telescopes that are heavily controlled by computers, let's look at the twin Keck telescopes. Here they are on the summit of Mauna Kea in Hawaii. The Keck telescopes are 10-meter reflecting telescopes, but 10 meters is larger than the largest piece of glass we can make for a mirror. How do they do that? Well, the Keck telescopes are actually a segmented mirror.

In this image, you can see the mirror stretching 10 meters across, and I've overlaid these hexagons to show you the segments of the mirror. There are actually 36 hexagonal segments that build up this entire mirror. Behind each mirror, there are censors and actuators, and twice a second, computers line up the edges between those segments. The edges have to be lined up to a very precise amount because you have to get down below the wavelength of light. They are lined up to 4 nanometers twice a second. Without computers, the Keck telescope would be impossible.

You can also use computers to link together telescopes, and this is very common in radio observations. Here we have a radio telescope array, and each of these dishes is a radio telescope in itself. But all of the radio telescopes in this long array are linked together by computers, and all of their observations are brought into the central computer and correlated together, and by combining the signals from all of them, you effectively get one radio dish as large as the entire array.

What's even more impressive is that you can take an array here in the United States, an array in Europe, an array in Australia—all around the world—and tie them together, and you effectively get a radio telescope as large as Earth. This is being done, and it's called Very Long Baseline Interferometry. They provide radio telescope observations that are the highest resolution that we can get.

As good as that is, it still helps to get off our planet and above our atmosphere because Earth's atmosphere actually blocks most of the light reaching it. As we can see in this diagram here, the visible light is transmitted through Earth's atmosphere, and radio waves mostly get through Earth's atmosphere. Infrared light can be seen pretty well from mountaintops, but most of the rest of the radiation doesn't get through our atmosphere. Most of it is absorbed, either high in the atmosphere or about halfway down through the atmosphere.

Now, this may sound strange because we use sunblock to protect us from the UV radiation of the Sun. Not that much UV radiation is truly getting through. If the complete amount of UV radiation were truly getting through, we'd need some really serious sunblock.

The other thing Earth's atmosphere does is, of course, distort the light that does come through. The observations we make from telescopes on the ground are a little bit blurry because of the motions within Earth's atmosphere. We can really only see resolutions down to about 1 arc second on most telescopes on the ground. So getting above Earth's atmosphere will enable us to see wavelengths that we can't see from the ground and will also give us better observations in the wavelengths we can see.

The most famous satellite telescope is the Hubble Space Telescope. Here it is, shown in its orbit above the Earth. It is about 360 miles above Earth, and if you look on this image, you can see along the Earth's atmosphere, the atmosphere is only about 60 miles high. So Hubble skims along just above Earth's atmosphere. We say that it's in "low Earth orbit."

I often describe Hubble as having a bird's-eye view because it's up high and it has very fine resolution. Hubble, not being limited by Earth's atmosphere, has a resolution down as fine as 1/20 of an arc second. This gives it the finest resolution in visible light of any telescope, and most of the spectacular nature of the images from Hubble comes from this vantage point in space.

Now, this is low Earth orbit. In contrast, the Spitzer Space Telescope is much farther away. Instead of being hundreds of miles from Earth, it is tens of millions of miles from Earth. It's actually farther from Earth than the Moon. Spitzer is an infrared telescope, which you may know is associated with heat radiation. And it is advantageous for Spitzer to be out at that distance, so that it gets away from the heat of Earth.

Hubble and Spitzer are just two of NASA's "Great Observatories." The other two are the Compton Gamma Ray Observatory that observes in gamma rays—that's already completed its mission—and the Chandra X-ray Observatory that observes in X-rays. These are a suite of observatories placed in space to do general observing in four major wavelength bands.

But we also sent a fleet of other more specialized observatories to look at the universe in many, many different wavelengths. In this chart, we show many of the space missions that have been launched and the wavelength coverage that they've got. We can see we have several missions out here in gamma rays, many in X-rays, some in UV, visible, through the infrared, microwave, and we even have some radio telescopes in space. These missions allow us to see the universe in new ways and in finer detail than we could from the ground.

Finally, beyond just looking at the universe in the space age, we have gone there. We have taken the next step and become explorers of our solar system. We have manned and satellite missions that have gone across the solar system and uncovered details that enrich and have transformed our scientific views. The most famous of these is, of course, the Apollo missions to the Moon.

These Apollo missions transformed the Moon into a real place—and transformed humanity into adventurers across the universe. Here's one of the most famous images. This is Buzz Aldrin and the American flag on the Moon. And it evokes a sort of patriotic sense that we have conquered—well, I don't think anyone thinks we've really conquered the Moon, but at least we have conquered the way to get to the Moon.

As an astronomer, there are two things I notice about this. Well, three things, actually. First is that the flag is being held up. All right? That waving in the flag, that's not real because the second thing I know about this is there's no atmosphere on the Moon. If you look beyond the horizon, it's pitch dark. The third thing I notice about it is all these footprints down here. And if you think about it, there is no atmosphere. There is no rain. There is no erosion on the Moon, and those footprints are still there. Those footprints will be there for billions of years.

Another thing, as an astronomer, that I like about the Apollo missions to the Moon is not necessarily the pictures of the astronauts, but they're the pictures of the Moon itself. This image of a crater, taken from the Apollo obiter, somehow speaks to me. It really brings up a visceral sense of the Moon, and you've got this giant crater and all these small craters around it. I immediately start imagining the impact events that must have formed this surface. To me, the close-up images that we have of the Moon and all the other planets from our space missions have taken these places and put them into my mind. They've provoked my imagination about what the universe is really like. They give, I don't know, *texture* to the solar system.

But even before we went to the Moon, we started visiting other planets. This image here is a drawing of the first interplanetary probe, Mariner 2. It was launched August 27, 1962, on a mission to Venus. It was followed by many other missions. They were the Venera probes to Venus, which by the way landed on its surface and then promptly melted because the temperatures on Venus are so hot. There were the Mariner missions to Mercury, the Viking missions to Mars, Galileo to Jupiter, Cassini to Saturn. We even have the New Horizons mission going out to Pluto, and it'll get there around 2015.

There are two sets of missions in particular that struck me as I was growing up. The first are the Voyager missions to the outer solar system. Shown here in this diagram are the plots of the Voyager missions. Voyager 1 went past Jupiter in '79 and Saturn in '80 and then continued on out of the solar system. Voyager 2 followed shortly thereafter and continued on to visit Uranus in 1986 and Neptune in 1989. They took some amazing pictures of these planets. They discovered moons and rings, and they transformed our ideas about the outer solar system.

It really ignited my imagination, and what's really nice about these Voyager missions is that they're still going. They have exited the inner part of the solar system, and now they're about 100 AU away from the Sun, and they're on what they call the Voyager Interstellar mission. In a few years, they're going to reach the edge of the solar wind and explore the heliopause. These two small satellites, launched in the late 1970s, are still going 30 years later.

The other missions that really excited my imagination have been this fleet of Mars missions, starting with Viking in the '70s and extending on through the several missions we have to Mars today.

We have orbited, we have landed, and we have explored the red planet.

This image here is one of the best images of Mars that we can get from Earth. It's an image taken by the Hubble Space Telescope, and it's fantastic for an Earth-based image. But if you really want something cool, well, we've been there. This is a famous image called *Twin Peaks*, referring to the two peaks in the background, and basically it looks like a rock garden on Mars. It was taken by one of the rovers, and the rovers have moved throughout the landscape and examined the geology and searched for water. But more powerful is the infusion of this as a place into my imagination.

We have transformed those tiny spots of light that I've seen through telescopes and made them into known worlds. We have begun to explore our neighborhood in space, and it has become part of our thinking. When historians write the history of the solar system, they tend to focus on the Copernican revolution. It's often cited as the biggest change in our view of the solar system, when we went from the Earth-centered to the Sun-centered.

As an astronomer, I have to disagree. I think that the space age is the more important epoch. The space age brought us new ways to observe the solar system in more wavelengths and with bigger telescopes; new ways to analyze with better data and faster processing; and also a new way to explore, both with robots and with our own eyes. We have gone through the breadth of the solar system.

This image here, a very famous image, *Earth Rise*, symbolizes the change in perspective. When we, as humans, see our own planet rising, and we see it in space as it truly is, we no longer think of the system as "out there." It has become real up here. It has become part of our mental model, and we see our place in the solar system anew.

The capabilities enabled by the space age have changed our perspective more than anything else in history. In the next several lectures, we will explore in detail the range of observations, ideas, and discoveries enabled by the space age. Until then, have a wonderful day.

Lecture Eight
The Star of Our Solar System

Scope:

Our Sun is a typical star, similar to a hundred billion others in our Milky Way Galaxy. However, since all other stars are at least 250 thousand times farther away, we can determine the Sun's characteristics to an unparalleled degree. Observations of the Sun's atmosphere in ultraviolet and X-ray light reveal a dynamic, high-energy environment that is in striking contrast to its somewhat placid visible surface. Theoretical models and a technique called helioseismology have helped determine the Sun's multilayered structure down to its intensely hot and dense core. Within that core, nuclear fusion converts enormous amounts of matter into energy in order to power the prodigious output of the Sun. Our detailed knowledge of the Sun forms the basis for understanding all other stars in the universe.

Outline

I. The Sun is in many ways an average star.
  A. While there are many billions of stars in our Milky Way Galaxy, the unaided human eye can see only about 6000 of the nearest and brightest.
    1. Most stars are in a stable phase of their lives called the main sequence.
    2. Stars are composed of about 75% hydrogen and 25% helium gas.
  B. The Sun is neither the biggest nor the smallest of stars in its mass, brightness (called luminosity), or temperature.
  C. The Sun is about 4.6 billion years into an expected lifetime of about 12 billion years.
  D. The Sun is metaphorically, geographically, gravitationally, and energetically the center of our solar system.

II. The atmosphere of the Sun includes the photosphere, the chromosphere, and the corona.
  A. The photosphere is the visible surface of the Sun.
    1. The temperature in the photosphere is about 5500°C.

2. Atoms of matter are comprised of a nucleus, containing protons and neutrons, surrounded by an electron cloud.
3. The solar gas is a plasma: atoms that have been excited to a high enough temperature to dissociate all of their electrons.
4. The visible light appearance of the Sun is relatively calm, with Earth-sized dark splotches called sunspots.
5. Sunspots are regions of intense magnetic fields that appear dark because they are cooler than their surroundings.
6. The photosphere is covered in millions of granules, the tops of columns in the convective layer beneath.

B. The chromosphere and transition region are thin, dynamic layers above the photosphere.
1. The chromosphere is about 1000 kilometers thick, and the transition region is about 100 kilometers thick.
2. The temperature would be expected to decrease in these regions, but instead it rises rapidly to about 1,000,000°.
3. Short-lived columns of high-temperature rising gas, called spicules, traverse the chromosphere.
4. Prominences and filaments are long streamers of cooler gas above the chromosphere that are seen on the limb (prominences) or the face (filaments) of the Sun.

C. The corona is a very hot, but very low-density, region extending out millions of kilometers from the Sun.
1. Gas temperatures in the corona are a few million degrees; they are heated by activity in the solar magnetic field.
2. Pressure in the corona is extremely low, about 1/10,000,000,000 of the pressure of Earth's atmosphere.
3. Whereas the photosphere and chromosphere are governed by gravity, the corona is dominated by magnetic fields.
4. The corona is constantly changing, with magnetic fields creating loops, holes, and outbursts.

III. The internal structure of the Sun contains three distinct layers. The core occupies the inner 20%, the radiative zone goes out to 70%, and the convective zone is the last 30%.
A. The internal structure of the Sun is governed by the principles of hydrostatic equilibrium and heat transfer.

1. Gravity is an attractive force that presses toward the center of the Sun.
2. Heat produces thermal pressure that pushes outward and counteracts gravity.
3. The transfer of heat can be done by conduction, convection, and radiation.
4. The three regions inside the Sun are defined by the way energy is transported across each of them.

B. Energy is generated in the core.
1. Temperatures in the core reach 15,000,000°.
2. The core contains about half of the Sun's mass.
3. All of the Sun's energy is generated in the core, and then that energy diffuses out through the overlying layers.

C. The radiative zone is a calm region where the Sun's energy is transported by photons.
1. The radiative zone is an example of hydrostatic equilibrium.
2. Energy takes tens of thousands of years to cross the radiative zone.
3. The temperature cools to about 2,000,000° across the radiative zone.

D. The convective zone is a turbulent region where movement of gas transports energy.
1. Large cells of heated gas rise to the surface where they cool by radiating in the photosphere.
2. The convecting material takes about 10 days to rise to the surface.
3. The temperature drops quickly near the top of the convective zone to 5500°C.
4. Even though the convective zone only occupies the outer 30% of the Sun's radius, it is about two-thirds of the Sun's volume.

E. Interior properties of the Sun can be measured with helioseismology.
1. Doppler shift is a change in wavelength caused by motion.
2. Sound waves oscillate throughout the Sun with millions of wavelengths and can be detected by measuring Doppler shifts on the surface.

3. Different wavelengths penetrate to different depths and reveal the Sun's internal structure and motions.
4. The sound waves are damped at the photosphere, but some leakage of energy is what powers the spicules in the chromosphere.

IV. The Sun is powered by nuclear fusion in its core.
   A. The extreme conditions in the core allow nuclear fusion.
      1. In the core, the temperature is 15,000,000°, the pressure is 200 million times that of Earth's atmosphere, and the density is 13 times that of lead.
      2. The proton-proton chain combines four hydrogen nuclei to form one helium nucleus and gives off energy.
   B. Nuclear fusion reactions in the Sun convert mass into energy.
      1. Einstein's famous equation, $E = mc^2$, is an expression of the equivalence of mass and energy.
      2. In the equation, E represents energy, m represents the rest mass, and c represents the speed of light.
      3. In fusing hydrogen to helium, 0.7% of the mass is converted to energy.
      4. The Sun converts mass to energy at a rate of 5 million tons every second.
      5. About 37% of the hydrogen in the core has been converted into helium.
   C. The neutrinos produced during fusion can be observed on Earth and provide evidence of the nuclear reactions in the Sun's core.

V. A star may be described as a gravitationally confined nuclear fusion reactor.
   A. The same physical processes we see in the Sun take place in all stars. The main sequence is defined by the presence of hydrogen fusion in the core of a star.
   B. Energy generated in the core slowly diffuses to the surface, where it is radiated into interplanetary space.
   C. The temperature of the Sun decreases from core to surface but then increases into the atmosphere.

D. The magnetic fields of the Sun, to be discussed in depth in a later lecture, generate enormous activity near the surface and in the atmosphere.

Suggested Readings:

Beatty, Petersen, and Chaikin, *The New Solar System*, chap. 3.

Bennett, Donahue, Schneider, and Voit, *The Cosmic Perspective*, chap. 14.

Hill and Carlowicz, *The Sun*.

McFadden, Weissman, and Johnson, *Encyclopedia of the Solar System*, chap. 4.

Shu, *The Physical Universe*, chaps. 5–6.

Wilson, *Astronomy through the Ages*, chap. 11.

Questions to Consider:

1. Before nuclear fusion was discovered, what would astronomers have thought powered the Sun? What important problems would those hypotheses have in terms of generating energy and sustaining output for billions of years?

2. What types of telescopes are best suited to studying the Sun? Why has the space age revolutionized solar research?

3. What do you think will happen when a main-sequence star starts to run out of hydrogen in its core?

Lecture Eight—Transcript
The Star of Our Solar System

How many stars can you see in the night sky? When you look up without a telescope, how many do you think you can see? Most folks will answer "millions of stars" because, intellectually, they know there really are millions and millions of stars out there. But if you actually went and looked, and tried to estimate, what would you find? Of course, it would depend upon where you were looking. When I was in New York City, we used to do a star party in Central Park, and really, we could only see a few hundred stars. I've also done star parties at Bryce Canyon National Park, and there were too many stars to count.

The truth is that the human eye can see about 6000 stars, and that's looking at stars all over the globe. Since you can really only see about half the sky at one time, you really can only see just a few thousand stars.

This is one such star field. It features the constellation Orion, and it has two very prominent giant stars. The shoulder of Orion, this red star here, is called Betelgeuse; it's a red supergiant star. The knee of Orion is Rigel; it's a blue supergiant star. These are stars in their old age. They're on their way to die. And of course, they're going to die by a very spectacular way, a supernova explosion.

On the left side of this image, this bright star over here is Sirius. It's the brightest star in the night sky, and it's not an old-age star. It's actually a middle-aged star. It's part of what we call the main sequence. These are stars that are producing energy in what we call the normal fashion. It's a stable phase that lasts, for many stars, billions of years.

Stars have a composition of about 75% hydrogen and 25% helium. The reason for this is simply those are the proportions made in the big bang. In the big bang, the nucleosynthesis, as it forges the elements, creates three-quarters hydrogen by mass, and one-quarter helium by mass. Stars in the main sequence are turning hydrogen into helium, and these are the majority of the stars that you will see in the night sky.

Our Sun is no exception. It is part of the main sequence, but it's rather an unexceptional main-sequence star. By mass, it's not the biggest—the biggest stars are about 100 times the mass of our Sun—

and it's not the smallest. Our Sun is about 14 times more massive than the smallest main-sequence stars.

In its energy output, or what we astronomers call luminosity, the brightest stars have a luminosity factor of 1 million greater than our Sun, and the faintest are about 1/1000 the luminosity of our Sun. The temperature range of stars goes from about 2000° in the low end to about 50,000° on the upper end.

Our Sun has a temperature of about 5500°. This is toward the lower end, and you know, you might be thinking, "Oh, our Sun is below average." But that's actually a good thing because the most massive stars have a motto of "live fast, die young, and leave a good-looking corpse." The corpse, of course, being those supernova remnants produced by those explosions that I talked about earlier.

These most massive stars will only live for about 50 million years, not long enough for life to develop on planets around them. Our Sun is about 4.6 billion years into an approximately 12-billion-year lifetime. And, as we are evidence, it's long enough for life to develop around it.

Now, the most important thing about the Sun is that it is *our* star. Every other star is more than 250,000 times further away than our Sun. That means that the Sun is the only one in which we can see detailed structure, and we use it as a template for understanding all stars in the universe.

For our solar system, the Sun can be thought of as the sine qua non, that without which it doesn't exist, because it contains 99.9% of the mass in the solar system. Everything revolves around the Sun, both metaphorically and literally, and the energy produced by the Sun enables life here on Earth. If something is that important, well, I think we ought to know something about it.

So, let's take a look at the structure of the Sun. We'll start with the outer layers, its atmosphere. There are three layers of the atmosphere that I will discuss: the photosphere, the chromosphere, and the corona. The photosphere is the visible surface of the Sun. Now, the Sun doesn't have a solid surface, but it does go through a very strong pressure drop. It goes from 10,000 times the atmospheric pressure of Earth down to 1/10,000 the atmospheric pressure of Earth over a very small radius of the Sun. And this is where the surface of the Sun is.

As I said before, it's about 5500°C. That is hot enough to melt any known substance. It is also hot enough that the gas becomes a plasma. Atoms are composed of an atomic nucleus and an electron cloud. And when you heat them up, the electrons become dissociated with the atoms. In a plasma, all of the electrons have become totally dissociated from all their atoms, and what you've got is a mixture of atomic nuclei and electrons not attached to any particular atomic nucleus, running all around. They're totally dissociated.

This is an image of the photosphere of the Sun, and you can see it looks relatively calm. It's a smooth yellow ball. You can also see that it has some freckles on it. These freckles are called sunspots. This is a close-up of a sunspot, and this has sort of the classic sunspot circular shape. It looks a little bit like the iris of an eye. Even a small sunspot is rather large on human scales. If I put it up next to our planet, you can see that even a small sunspot is about the size of our entire planet.

Some sunspots are much, much larger. Here's a sunspot group that shows a wonderful spiral pattern. It's rather rare to find one with a spiral pattern. And it is many times larger than our planet. These sunspots are islands of intense magnetic fields. They appear dark, not because they are not shining, but just because they are slightly cooler than the surrounding photosphere. They're still very, very bright. It is just the contrast between the much brighter region around them and the material inside the sunspot that makes them appear dark.

These sunspot groups can last for months. Much smaller and also shorter lived are granules, which are fine structures on the surface of the photosphere. If I look in closer to the Sun, you'll see a pattern like this. These individual cells are the granules, and there are millions of them across the surface of the Sun.

Each of these granules is approximately 1000 kilometers across, and they are the tops of columns of material, moving up from the interior of the Sun, radiating their energy away, and then flowing back down. Each of these granules lasts for only about 10 minutes. So there are millions of them. They regenerate themselves on the order of 10 minutes. The surface of the Sun is actually a very active place.

Above the photosphere is the chromosphere. Traditionally, the chromosphere was a colored portion of the Sun that you could see during a total solar eclipse. Here's an example of that. This is a total solar eclipse, when the Moon has blocked out the full disc of the Sun. And do you see that tiny little red region, just along the edge of the Moon? That was traditionally identified as the chromosphere.

But our modern observations have sort of redefined the chromosphere. We have the chromosphere that extends for about 1000 kilometers above the surface of the Sun. We also have something we call the transition region, which extends for another 100 kilometers above that. In this region, the temperature of the Sun rises dramatically. It goes from 10,000° to 1,000,000°. And this is absolutely crazy because if the energy is being released from the Sun at the photosphere, everything should be cooling down. If the energy is gone here, everything should be cooling down as you get further out. Instead, the energy is increasing.

Where is this energy coming from? Well, the answer that we've found so far is that it's coming from a structure called spicules. This is an image of some spicules on the surface of the Sun, and it's taken in ultraviolet radiation. You can see these very thin jets of gas. They only last for about 5 minutes, and the energy is being transported from the photosphere, up through the chromosphere and transition region, to the corona. The energy is leaking out from the interior. It's creating a shock wave in the chromosphere, and it's driving these spicules.

There's a wonderful little movie of these spicules. This was taken by the STEREO spacecraft, and in it, you can see these very fine spicules flitting in and flitting out. This is an image taken every 10 minutes for about one day of spicule activity. It's very high-temperature gas that's shooting up through the chromosphere.

But not all gas above the chromosphere is hot. There are also some streamers of cooler gas that appear, and in this image taken in hydrogen-alpha, a specific wavelength emitted by hydrogen atoms—and you can see this very long, dark thing here. This is called a filament, and if I zoom in on it, what this filament is is gas that is cooler than the material behind it, and shows up dark. And it's being held aloft above the photosphere by magnetic loops. And these filaments can last above the chromosphere for several weeks.

They appear dark here when they're viewed against the Sun, but if they are seen on a limb, as in this image, you can see they are actually quite bright. These are called prominences. Prominences and filaments are really the same phenomenon. One is seen against the disc of the Sun and appears dark, and one is seen against the background sky and appears bright.

The third region of the Sun's atmosphere is called the corona. This is a very hot, very low-density region, and its temperature is a few million degrees. We really only see the corona well in ultraviolet and X-ray emissions. Its pressure is extremely low. It's down to 1/10,000,000,000 of Earth's atmospheric pressure. We can see the corona with our own eye during total solar eclipses.

This is an image of a total solar eclipse, and you can see all these streamers, coming out on either side of the Sun, are part of the corona. It's a rather wispy region, and that extends to several solar radii. The material here is flowing outward and eventually becomes the solar wind. The physics of the corona is dominated by magnetic fields, and that's in contrast to the photosphere and the chromosphere, where gravity dominates. When you saw those spicules created, they fell back due to gravity. Everything in the corona is generally held up by magnetic fields or pushed out by magnetic fields.

You can see that in this image, where I show you coronal loops, taken in the ultraviolet. And the plasma here, incredibly heated plasma, is arcing high into the corona. It's heated at the base and streaming up through the magnetic field, back down to the Sun, through about half a million kilometers. The corona is also the site of many storms and eruptions, but we're not going to discuss those today. Those are part of a later lecture.

So now, we're going to turn inward and look at the internal structure of the Sun. The internal structure of the Sun also has three layers. There's the core, there's the radiative zone, and there's the convective zone. Each layer is defined by how it transports energy. To understand this a little bit, you have to get down the concept of hydrostatic equilibrium, and it's not a difficult concept. It's really just the balance of forces.

Gravity is the major force pulling down on the Sun. The Sun is incredibly massive, and all that mass presses down towards the center and creates a huge force pushing towards the core. What pushes back? Well, thermal pressure. When you heat something up, it creates a pressure. You may remember the ideal gas law from your chemistry courses in high school. Increase the temperature, you increase the pressure. This is the balance going on in the interior of the Sun: gravity pushing down, thermal pressure pushing out.

We also have to think about heat transfer, and there are three ways that you transfer heat: conduction, radiation, and convection. And you can understand all of those by thinking about your stove. If you've ever touched the burner on the stove and gotten a burn, that's conduction, when heat flows directly from one substance to the next. If you just hold your hand above the burner, and you can feel the heat radiating off of it, that's radiation. If instead, you put a pot of water on your stove and bring it to just before a boil, where you can see the water roiling and rolling, it's heated at the bottom, rises to the top, and releases that heat—that's convection.

So, this is the internal structure of the Sun. It's basically very simple. The core occupies the inner 20%, the radiative zone goes out to 70%, and the convective zone is the last 30%. In the core is where the Sun does all energy generation. It's at a temperature of 15,000,000°. It contains about half the mass of the Sun in that small core. It creates all the energy that then diffuses out through the other zones of the Sun.

The radiative zone is a relatively calm region because the energy is transported by photons. It is in hydrostatic equilibrium, so material isn't moving back and forth for the radiative zone, just the energy is flowing through the radiative zone. And that energy takes tens of thousands of years to cross the radiative zone. Why?

Well, because the photons are doing what we call a random walk. The photons that are emitted from the core travel out, and they are quickly absorbed. And then they are reemitted in a random direction, and they're quickly absorbed. And they're reemitted and absorbed, and reemitted and absorbed, and reemitted and absorbed. And they go off in these random directions and slowly, slowly diffuse through the radiative zone.

I've seen estimates that say it takes between 30,000 years to 170,000 years for a photon to travel across the radiative zone. In the radiative zone, the temperature starts out at the 15,000,000° of the core and cools down to about 2,000,000°.

On top of the radiative zone is the convective zone, and this, in contrast, is a very turbulent region. Here is where movement of the gas transports the energy. At the top of the radiative zone, the gas heats up. That heated gas rises to the surface of the Sun, it cools by radiating the photosphere, and then cycles back down. It takes about 10 days to rise to the surface, quite a contrast to the tens of thousands of years it takes to cross the radiative zone. You get across the convective zone in a little over a week. The convective zone is filled with large convective cells, but this is not what we see in the photosphere. At the photosphere, we see the tops of much smaller columns. These are the granules that I showed you earlier.

In the convective zone, the temperature of the Sun drops strongly. Here is a plot of the temperature profile of the Sun's interior. And starting on the right, you can see it's very hot, the 15,000,000° in the core. It cools slowly across the radiative zone, and then it cools very, very quickly down to the 5500° in the convective zone.

This shows you the efficiency of the convection. It's a very efficient process for getting rid of heat. It also shows you that the convective zone is actually quite large. Even though it is only the outer 30% of the Sun, it contains about two-thirds of the Sun's volume. That helps it be a very efficient way of transporting heat.

Now, you might be thinking to yourself: "That's a nice story. Sure you know exactly what's going on inside the Sun. Do you have any proof of it?" And previously, I would have had to wave my hands and say, "Well, we've got these wonderful hydrostatic equilibrium models, all this theoretical stuff." But I don't have to any more.

In the last decade or two, we have a technique called helioseismology that allows us to look into the interior of the Sun. Now, you probably have encountered Doppler shift. Doppler shift is a change in wavelength due to motion. You're familiar with it when a car goes by, and it has a high-pitched sound as it's approaching and a low-pitched sound as it's receding. That's Doppler shift because the motion changes the sound of the car as it goes past you. As it's approaching you, it goes up. As it's going away from you, it goes down.

Those same sort of acoustical waves, sound waves, occur inside the Sun. They oscillate the interior of the Sun, and they cause the surface to pulsate. We can measure those pulsations by seeing Doppler shift in the spectral lines of the Sun. If you look at those spectral lines, those spectral lines will pulsate back and forth with the Doppler motion of these sound waves.

Here's an image of a single sound-wave mode. This is a mode that is pulsating throughout the entire Sun, and the receding portions of the Sun are in red, and the approaching parts are in blue. And this wave here oscillates on a period of about 5 minutes. There are millions of other waves that can be detected by a group known as the Global Oscillation Network Group. This is a collaboration of solar observatories spaced evenly across the globe so that one of them is always in daylight and is always studying the Sun.

We have a large helioseismology consortium that is continuously studying the Sun and determining all of these oscillation periods. As I said, there are millions of wavelengths that they have found, each with a different size scale and different period. And when you've got different wavelengths, they will penetrate to different depths within the Sun, and the varying strength of those wavelengths will enable us to determine the Sun's internal structure.

So, we are extremely confident about that story I just told you about the Sun's internal structure because, finally, we have actually measured it. One little note: The sound waves that we talk about in this helioseismology, they're usually damped at the photosphere, and you need that in order for oscillation. You need to have a harmonic between two points, but sometimes those sound waves leak through, and that is the leakage of energy that powers the spicules.

But the real energy of the Sun is down in the core, and that's where nuclear fusion occurs. There are extreme conditions in the core. As I said, the temperature is 15,000,000°. The pressure is 200 million times that of Earth's atmosphere. The density is 13 times that of lead. In those extreme conditions, you could have nuclear fusion.

Nuclear fusion goes by a process we call the proton-proton chain. In this diagram, we show you the basics of the proton-proton chain. First, two protons, or you may want to think of them as hydrogen nuclei, join together and become a deuterium nucleus. Then that deuterium nucleus is smashed by another proton, and you get a helium-3 nucleus. When two helium-3 nuclei smash together, you get a helium-4 nucleus, or normal helium nucleus, and two protons come off.

The basic result is that four protons go into the reaction, and one helium nucleus comes out. And in that process, occurring at these incredible temperatures and incredible pressures, you can convert mass to energy. This comes through at the famous equation $E = mc^2$, where E is energy, m is mass, and c is speed of light, squared. You see, the mass of the four protons going in and the mass of the helium nucleus coming out are slightly different. The protons weigh slightly more than the helium nucleus.

The difference in mass is converted into energy via $E = mc^2$, where 0.7% of the mass going into the reaction comes out as pure energy. The Sun shines because it is turning mass into energy at a rate of 5 million tons every second. I'm not talking about 5 million tons of protons. I'm talking 5 million tons of that 0.7% that's being removed from the universe every single second. Now, that's going to scare you. Wait a minute. Is the Sun going to last long with that rate of energy burning? Don't worry. The Sun is about a billion billion billion tons, so it's going to last for a little bit longer. Matter of fact, we estimate that about 37% of the hydrogen in the core has been converted to helium by this time.

In the nuclear process, you not only create energy, you also give off these ghostly particles called neutrinos. Neutrinos are particles that don't interact with just about anything. There are, right now, probably a billion neutrinos passing through my body, and I don't know it. They don't interact with matter at all. But sometimes you can get them to interact. We have neutrinos telescopes, and the original neutrino telescope was basically a swimming pool filled with chlorine. Once in a great while, about once per day, a neutrino would interact with a chlorine atom and turn it into an argon atom.

And to use this neutrino telescope, what they did is they filtered the chlorine and searched for those elusive atoms of argon, indicating an interaction with a neutrino, and in that manner were able to measure neutrino flux from the Sun. We have measured it, and we do have evidence of the nuclear fusion processes going on in the core of the Sun.

So, that is the structure of our star, and it is also the structure of every other star. A star may be described as a gravitationally confined nuclear fusion reactor. Gravity provides the force that pushes down. Nuclear fusion provides the temperature that creates the pressure that pushes back. This is the same physical process that occurs in all stars. Matter of fact, the main sequence is defined by stars undergoing hydrogen fusion.

The energy is generated in the core, and it very slowly diffuses through the radiative zone, and then it's relatively quickly transported through the convective zone. It's radiated from the photosphere, and it takes about eight minutes to reach Earth. The timescale from when the energy is generated to when we see it on Earth could be as long as about 170,000 years. And if that's true, I've got an interesting thought for you.

This image is of the Small Magellanic Cloud. It is a satellite galaxy of our Milky Way Galaxy, and its distance is 170,000 light-years away. That means the light from the Small Magellanic Cloud has taken 170,000 years to reach us. That is the same amount of time that the energy has taken in coming to us from the Sun. The energy from our nearest star reaches us in the same amount of time as the light from one of our satellite galaxies.

The temperature in the Sun decreases from the core to the surface and then increases into the atmosphere. The magnetic fields of the Sun provide enormous activity that drive that temperature increase, and even more in the atmosphere—create explosions and huge coronal mass ejections. But we're not going to get into those today. We'll get those in a later lecture on solar storms.

And we'll just leave the story of the Sun there for now. We'll see you next time. Take care.

Lecture Nine
Planetary Surfaces and Natural Wonders

Scope:

The natural wonders of the world are magnificent surface features of Earth. Yet almost all of them are duplicated and exceeded elsewhere in the solar system. Our highest mountains are surpassed and even dwarfed by those of our neighbors. Our deepest canyons can seem but a minor scar by comparison. Our fiercest volcanic activity appears relatively tame. Many similar features are found on other planets and moons, albeit with intriguing variations. From an interplanetary perspective, it is Earth's oceans that are its signature characteristic—a characteristic that is directly related to its primary unique feature: life.

Outline

I. The study of other places in the solar system is best viewed in the context of comparative planetology.
   A. We will look at the rocky planets and the larger moons.
   B. We learn about other worlds by comparing their features to those on Earth.
   C. We learn about Earth by seeing the range of possibilities and where our planet fits in.

II. A tour of rocky surface features reveals many natural wonders of the solar system.
   A. Mountains are found on several worlds but are formed in different ways on each.
      1. Earth features long chains of mountains that are the result of plate tectonics.
      2. Mountains exist on the Moon; some of them are similar in height to those on Earth and are the result of impacts.
      3. Venus has several large mountain chains, probably formed by compressional tectonics.
   B. Volcanic activity is abundant in the solar system.
      1. Volcanoes on Earth form when oceanic plates meet continental or other oceanic plates.

2. Venus has more than 1000 volcanoes, and about 85% of its surface is covered by volcanic plains.
3. Venus has a unique volcanic feature called coronae, which are oval-shaped volcanic uplifts.
4. Mars has several very large shield volcanoes, including Olympus Mons—by far the largest volcano in the solar system.
5. Io, a small moon of Jupiter's, is covered with volcanoes that are constantly erupting.

C. Canyons have been carved both by water and by geologic forces.
1. The Grand Canyon of the Colorado River on Earth is about 150 kilometers long and about 2 kilometers deep.
2. Venus features deep, narrow canyons called chasmata (singular: chasma) that are hundreds of kilometers long and show abundant geologic faulting.
3. Valles Marineris on Mars is up to 7 kilometers deep and roughly 4000 miles long. Its walls show attributes of both fault scarps and water erosion.

III. Hydrology on Earth is based on water, but elsewhere it shows remarkable diversity.
A. Features resembling riverbeds provide evidence of flowing liquids on many worlds.
1. Rivers on Earth create distinctive and recognizable patterns.
2. Mars shows ample evidence of water flows in the past, although liquid water cannot exist on its surface today.
3. Saturn's giant moon Titan shows river channels probably carved by liquid methane, as well as strong evidence of methane lakes.
4. Sinuous channels on Mars and the Moon were created by lava flows.

B. Geysers in the solar system spew a variety of substances.
1. Earth's geysers release heated water from beneath its surface.
2. Towering methane geysers have been observed on Neptune's large moon Triton.
3. Jets of icy particles spewing from Saturn's moon Enceladus contribute to one of its rings.

- C. Polar ice caps on Earth and Mars grow and shrink with the seasons.
    1. Large portions of Earth's polar ice caps melt and refreeze each year.
    2. The polar caps of Mars can nearly vanish in the summer as the carbon dioxide sublimes into the atmosphere.
- D. Earth's most distinguishing feature is the oceans that cover nearly 70% of its surface.

IV. While it is natural to marvel at the diversity of surface features, one learns more by identifying their similarities.
- A. Each of the rocky planets shows a dichotomy between rougher upland regions and smoother lowland plains.
- B. Similar features indicate that similar geologic and hydrologic forces are at work.

Suggested Readings:

Beatty, Petersen, and Chaikin, *The New Solar System*, chaps. 7–12, 17–22.

Bell, *Postcards from Mars*.

Bennett, Donahue, Schneider, and Voit, *The Cosmic Perspective*, chap. 9.

Benson, *Beyond: Visions of Interplanetary Probes*.

De Pater and Lissauer, *Planetary Sciences*, chap. 5.

Hey, *Solar System*.

McFadden, Weissman, and Johnson, *Encyclopedia of the Solar System*, chaps. 6, 8, 10, 12, 16, 22–26, 44.

Questions to Consider:

1. How is it that Mars, a smaller planet, could have a volcano 10 times taller than any on Earth?
2. Would you expect all of the material in the volcanic plumes above Io to fall back to its surface? If not, what would happen to the excess material?
3. Could biology have developed in the methane lakes of Titan?

Lecture Nine—Transcript
Planetary Surfaces and Natural Wonders

Hi. In this lecture, we're going to take a tour of the surfaces of the worlds in the solar system. Our focus will be on the rocky planets, but we'll also visit many of the larger moons. We'll do this in the context of comparative planetology, and in comparative planetology, we learn about other worlds by comparing their features to those of Earth. But we also learn about Earth by exploring the range of possibilities available in the solar system, and where our planet fits in.

Let's just jump right in. We'll start with mountains. Mountains are found on several worlds, and each forms in a different way. Here on Earth, if we look at the global map, we can see several very long mountain chains, such as along the West Coast of the United States in North America, and along the west coast of South America. If we look closer at South America, we can see the very long chain of the Andes Mountains, running all along its west coast. This is a result of plate tectonics, where the oceanic plate of the Pacific has run into the continental plate of South America.

This is not the highest mountain chain in the world. The highest mountain chain in the world is the Himalayas, and they're pictured here, north of India. This is where two continental plates have collided, the Indian Plate moving northward, smashing into the Asiatic Plate, and lifting up the Himalayas. The highest mountain in the world is in the Himalayas, and it's Mount Everest. And here is Mount Everest rising high over the Tibetan Plateau, and it reaches 28,000 feet high.

Mountains are not only on Earth, but they also exist on other worlds, and here's an example of a mountain on the Moon. It looks somewhat like that picture of Mount Everest, but this is Mons Hadley Delta, and this is the Apollo 15 mission. Mons Hadley Delta rises about 15,000 feet above the lunar marae, where the astronauts landed. So, mountains on the Moon are similar in height to those of Earth.

Matter of fact, there's even a mountain chain. If we back off a bit, we can see, running right across the center of this image, the Apennine chain of mountains on the Moon, or the "Alps" of the Moon. These are probably the mountains that Galileo saw when he first turned his telescope to the heavens and was able to see that not all of the bodies of the heavens were perfect spheres. He could see the mountains on the Moon, and that violated the principles of Aristotelian philosophy.

However, if you look even further out, to the full Moon, you can see that these mountains are not the result of plate tectonics. These mountains are actually just part of the rim of a very large crater. The mountains on the Moon are the result of impacts.

On Venus, we also have mountains, and this is a topographic view of Venus from its north pole, where blue represents the lowland regions and white represents the high regions. It's rather easy to pick out the highest region. It's down here. And if we zoom in on it, it's called Lakshmi Planum, and the white region here is Maxwell Montes. Maxwell Montes is the highest point on Venus. It's the tallest mountain on Venus. It is also, coincidentally, the only feature on Venus that is named for a man. It's named after James Clerk Maxwell, the man who derived the equations of electricity and magnetism, but every other feature on Venus is named for a woman, as you might expect.

Maxwell Montes also did not form by plate tectonics, and it's not the result of impact. It's rather the result of compressional tectonics. Venus is a single-plate planet. Instead of having multiple plates like Earth, it's one plate over its entire surface. But compression within that plate is what probably caused Maxwell Montes to rise up so high over the surface.

Next, let's explore volcanoes. This activity is abundant in the solar system; and on Earth, the most common form of volcano formation is when an oceanic plate hits another plate and subducts underneath it. The pressure created in the subduction zone sends magma to the surface and creates a volcano.

This volcano here is very recognizable: It's Mount Saint Helens. It is caused by the subduction of the Juan de Fuca Plate underneath the North American Plate. This image here shows Mount Saint Helens as it was before its 1980 eruption, and this image shows it after its eruption. That is the definition of "blowing your top."

These volcanic explosions can be incredibly powerful. There are maybe a few thousand volcanoes across the surface of the globe, and they are generally located on these plate boundaries. When we think of volcanoes, we often think of the Ring of Fire around the Pacific

Ocean, where the Pacific Plate meets the continental plates on either side.

Volcanoes are also observed on Venus. This is Maat Mons, and it's one of about 1000 volcanoes on Venus. I note that there is a ton of volcanic activity on Venus because 85% of its surface is covered by volcanic plains. Venus also features a unique volcanic feature that appears nowhere else on any other planet. This is Bahet corona, and it's an oval-shaped volcanic uplift. Apparently there are these volcanic uplifts that take entire regions of Venus's surface up and then compress them back down. And you can see all these fractures that form this oval-shaped region of a corona, and this corona is about 200 kilometers across.

We get much more volcanic activity on Mars. Here is one of the most volcanically active regions of Mars. It's the Tharsis uplift. There are several very large shield volcanoes, three down here, and the largest one up here. Shield volcanoes are created through very fluid lava that has been spewed out over a very long period of time. That's why these volcanoes are as large as they are.

"How large?" you may ask. Well, this large one up here is called Olympus Mons, and here's a close-up of it. Olympus Mons is by far the largest volcano in the solar system. This volcano by itself is about the same size as the state of Arizona. It's about 600 kilometers across. In height, it's about 27 kilometers high. And if I compare it to Mount Everest, the comparison looks like this. Olympus Mons basically dwarfs Mount Everest. It really does deserve the name "Mount Olympus."

But Mars is not the most exciting place for volcanic activity in the solar system. That claim goes to a large moon of Jupiter, Io. This is what Io looks like on the surface, and all of these pockmarks—I don't know, they look like pimples or such—they're actually volcanic regions. It's literally covered with volcanoes, and they are continually erupting. When the Galileo probe flew past it in April 2007, the region around the Pele volcano looked like it did on the left, and when it came past it again in September 2007, an entirely new volcano had erupted.

This incredible amount of volcanic activity is due to Io's elliptical orbit and Jupiter's gravity. As Io comes closer to Jupiter, its surface gets pulled in. As it goes farther away, it flexes out. And this flexion in and out of Io's surface creates the heat that causes the volcanic activity. Now, normally, this process should dissipate. The process of surface flexion should help circularize the orbit. But Io's in a resonance orbit with the other Galilean moons, Europa and Ganymede, and they keep Io on its orbit and maintain Io's volcanic activity.

One of my favorite images in all of astronomy is this, because the plumes of Io's volcanoes stretch so far up above its surface that you can see them on the limb. And here's a close-up of it. It's just amazing to me to think on Earth, to get a plume that high, you would have to be going, what, 300 or 400 kilometers into the air, and our volcanoes don't do anything like that.

Now, this was one of my ultimate favorite images, but then the New Horizons missions went past Jupiter, and it got this sequence of five images of the Tvashtar volcano on Io. And you can see five images that show it erupting, and the plume spreading out across Io. It's just an amazing sequence of images.

Next, we'll turn to canyons. Canyons are carved both by water and by geologic forces. On Earth, the most impressive canyon is in Arizona. Here is the view of the state of Arizona from space, and if we look in the upper region, we can see the Grand Canyon. It stretches from the upper right, along down, through the center, it wiggles around, and continues off to the left on this image. The Grand Canyon is about 150 kilometers long and about 2 kilometers deep. And it's been carved by the Colorado River over millions of years.

Now, Venus has canyons as well, and they're called chasma, or plural, chasmata. In this image, all of the bright yellow regions across the center and curving over here, around on the right, are these chasmata. They are very deep, very narrow canyons that stretch for hundreds of kilometers. They are not caused by water erosion. Instead, they're caused by geologic faulting.

On Mars, we have perhaps the most impressive canyon in the solar system. This is an image of Valles Marineris. It is a very extensive canyon system on Mars, and it shows both fault scarps and water erosion, so it has both of the processes that we see on Earth and Venus occurring on Mars. This canyon is so large that it can be seen when we take a picture of the entire globe. Here is Mars, and here is

Valles Marineris. It is roughly 4000 miles long. That is so long that if I compared it to the United States, it would stretch from California to New York. The canyon is about 7 kilometers deep and makes for a spectacular feature.

We have a wonderful movie that can take us on a fantasy trip through Valles Marineris. This movie is based upon topographic data from the Mars orbiter. We begin on the western edge of Valles Marineris, and then fly down the main canyon. The canyon was originally created by earthquake faulting many billions of years ago. As we come to the central region, a region called Melas Chasma, you can see a lot of sediments there on the right. I imagine an old ancient lake. As we pass to the north, we pass over an ancient landslide and into another basin. Notice the atmospheric effects on the Sun. They're rendered realistically, so that it looks what it would look like on Mars.

One thing that is not realistic is the vertical exaggeration. This is the same enhanced relief that you would find on a globe that has relief on it. As we pass to the northernmost wall of the canyon, you can see large landslides and examples of water erosion that have widened the valley over time. This really is a magnificent structure, extending for hundreds and hundreds of kilometers across the face of Mars. Now, they say they do things big in Texas, but considering Olympus Mons and Valles Marineris, Texas is nothing compared to Mars.

Next, let's take a look at riverbeds. Riverbeds provide existence of flowing liquids on many worlds. On Earth, rivers provide distinctive and recognizable patterns. We've already seen one example with the Colorado River in the Grand Canyon. We also see images that look something like this. These are dendritic channels and are a drainage pattern coming from a river channel.

It's very similar to the same features that we see on Mars, where we can see the same dendritic-like channels, and they show evidence of water flow on Mars. Not water flow currently, but water flow in the past. And Valles Marineris, as you follow it down through to its end, you can see that water once flowed along it. There are many other features on Mars that show water flowing in the past, as well as catastrophic releases—miniature floods—on Mars. But liquid water cannot exist on Mars's surface today. Its atmosphere is too thin, and there's just simply not enough pressure to hold water in its liquid form.

One place that's very surprising where we see these same riverbed channels is Saturn's moon Titan. These are methane channels on Titan, channels carved by liquid methane. We have also seen methane lakes on the surface of Titan, although you wouldn't want to go swimming in these lakes because methane is liquid at a temperature near −200°C.

There are other riverbed-like features that appear on other planets. Here is one on Venus, and we call it a sinuous channel. It looks like a river, but it's really been created by a lava flow. There is another one called the Hadley Rille on the Moon. The Hadley Rille was observed by the Apollo 15 astronauts, and matter of fact, if you follow the Rille to the north, right about here, around the bend, the Apollo 15 landing site was up in here. And this mountain right here is Mons Hadley Delta that we showed you before. The astronauts took the rover out to explore the Hadley Rille, and you can see what it looks like from the ground. Basically what it is, it's a collapsed lava tube that once flowed across the Moon.

The next topic to take a look at is geysers. They can spew a variety of substances. On Earth, the most famous geyser is this one, Old Faithful in Yellowstone National Park. It erupts for a few minutes on average every hour and a half, and it is one of about 300 geysers in Yellowstone Park. Matter of fact, nearly half of all the geysers in the world are in Yellowstone Park.

Geysers are geologic features that release heated water from beneath the surface. Ground water makes its way down through the rocks of the crust, reaches the magma underneath it, heats up, and is periodically released. These are rare features that exist only in very specific spots on our planet.

But we also see geyser-like features elsewhere in the solar system. This is Neptune's largest moon, Triton. And on the upper half of this image, you see what they call the cantaloupe terrain. Triton is extremely cold. It's out at the edge of the solar system, and its surface is covered with ice. In the bottom half, you can see streaks of black. Let me zoom in on these.

These streaks of black look like they're all streaming away from the south pole of Triton. Triton does have a very, very thin atmosphere, and we have determined these black streaks are methane fountains, liquid methane spewing up from the interior of Triton and being blown back by very, very thin winds, moving from Triton's south pole away. These methane fountains rise nearly 8 kilometers into the air before being streamed back. They're carried back by the thermal currents.

There is another moon that has even more impressive fountains. This is Saturn's moon Enceladus. Now, Enceladus is a rather small, icy moon, not a particularly important moon, except for these fountains. It orbits about twice the distance from Saturn as the main rings of Saturn, so the rings end about here, and Enceladus is way out here.

The important surface features of Enceladus are these cracks in the south. They call them the tiger stripes. They're very deep crevices into the surface of Enceladus, and for some reason, they seem to be heated. We're not exactly sure why they are heated, but they do seem to have excess heat associated with these tiger stripes. That extra heat appears to power some ice fountains.

Here is an image of Enceladus that is backlit, and down at the bottom here, you can see just these wispy fountains coming off the bottom part of Enceladus. Perhaps these tiger stripes are heated by tidal heating like Io, but we're really not sure. The interesting thing is this may indicate the presence of liquid water just below the surface of Enceladus. Again, we're not sure. The one thing we are sure of is that these fountains contribute to and may even be the main source of the faint E ring of Saturn. Saturn has some very bright main rings, but it also has some very faint rings on the outer part of it, and Enceladus may be the main source of one of its outermost rings.

Next thing to look at: polar ice caps. These exist on both Earth and Mars, and they grow and shrink with the seasons. Now, Earth has ice-covered poles, simply because they receive less sunlight; it's colder at the poles. Earth also has a 23.5° tilt, so in some months, the poles get no sunlight, and in some months, they get continuous sunlight. It makes sense that the ice caps would shrink in the summer and grow in the winter.

This is what the northern ice cap on Earth looks like in March of 1998. The ice cap is largest right after northern winter. And here is what it looks like in September 1998. It is smallest right after northern summer, quite a change between largest and smallest.

The same thing occurs on Mars. In this Hubble image of Mars, you can see that Mars also has polar caps, but these polar caps, instead of being water ice, are mostly carbon dioxide ice. Mars's atmosphere is 97% carbon dioxide. And, as we said, you can't have liquid water on the surface. Mars similarly has a tilt of 25°, very much the same type of tilt that Earth has, so you would expect to see seasonal changes in its ice cap, and we do see them.

Here is Mars over the course of many months. In the Martian winter, you can see it has a very large polar ice cap, and as it progresses towards the Martian summer, the polar ice cap shrinks. But this shrinking is not melting, as it is on Earth. On Earth, the polar ice caps are melting into the polar ocean.

Here, it's a process called sublimation. The carbon dioxide ice in the polar caps of Mars are going from ice directly into gas, and that gas is joining the Martian atmosphere. It's actually kind of interesting because we see the polar cap shrink on one pole of Mars. The ice goes into the atmosphere and actually freezes back out at the other pole of Mars.

We're going to finish off this planetary comparison with somewhat of a noncomparison because Earth does have one unique feature, and if you just look at it, you see it. It stares you straight in the face. It is the oceans. Oceans cover nearly 70% of Earth's surface. The important idea about the oceans is that water is a necessary ingredient for life as we know it. When we look for life in the universe, we are looking essentially for the presence of water. Life as we know it requires carbon chemistry, water, and a source of energy. Well, carbon chemistry and a source of energy are relatively abundant throughout the solar system, but water is the main ingredient we're looking for.

Now, it's interesting to note that only recently have we found another planet in the solar system, another world in the solar system that has liquid on its surface—and that is Titan, and it has the methane lakes. So while Earth has been unique for having water on its surface, it is no longer unique for having a liquid on its surface.

Let's take a look at a global overview of the rocky planets, take a look at their overall surfaces and compare them. Each of the rocky planets shows a dichotomy. It has rougher upland regions and smoother lowland plains. Earth and Venus are very similar. We look at this diagram. We see Venus on top and Earth down bottom. They are dominated by the blue color. That represents volcanic plains.

Venus has had many epochs of volcanic resurfacing, and its surface is almost entirely covered by volcanic plains. Earth has been dominated by plate tectonics, and you can see the dichotomy between the ocean bottoms, which are volcanic plains, and the continental plates.

You also see on Earth and Venus the pink regions that represent mountains. Venus has rather strange-shaped mountains compared to Earth because we are used to long mountain chains that you do not see on Venus. Venus's mountains would occur because of compressional tectonics, not because of plate tectonics.

You also see a unique feature on Venus: these very small orange circles. Those are the coronae that we mentioned earlier. On Earth, the colors that you don't see on Venus are green and tan. The green represents deposits on top of the continental plates, and the tan represents cratered regions. Although we have a lot of erosion on Earth, we still have a few regions that have significantly more craters. And this, of course, occurs on the continental plates.

When you look at Mars and the Moon, you see volcanic plains, the blue volcanic plains that we saw on Venus and Earth. It's very small on the Moon but really dominant in the northern half of Mars. The dominant feature on both of these planets is the tan that represent the cratered regions, as well as the yellow, which represent the heavily cratered regions.

On the Moon, we also see red regions that represent the ejecta. The Moon is a geologically inactive body, and almost all of the surface features you see here are due to the heavy bombardment era and the cratering that has occurred since then. On Mars, we see a very strong dichotomy between the volcanic plains in the north and the heavily cratered regions in the south. We don't know why this strong dichotomy exists, and it's something that the planetary scientists are studying.

It is natural to marvel at the diversity of surface features across the solar system. We have looked at mountains, volcanoes, canyons, riverbeds, geysers, ice caps, and oceans. And in looking at them, we have seen that the seven wonders of the solar system far outpace the seven natural wonders here on Earth. But I think one learns more by identifying the similarities rather than looking at the differences, because when you look at similar features you recognize that similar forces are at work, similar geologic and hydrologic forces are at work across the solar system.

Now, if you think back to what we just discussed on the Moon and Mars, and something that doesn't really appear that strongly on Earth, we've skipped something. What was it? Well, the Moon and Mars are covered with craters. It doesn't appear so strongly on Earth, but it's a very important surface feature across the solar system.

And they are so important and so prevalent that they get their own lecture. And we'll discuss them next time. See you then.

Lecture Ten
Craters, Impacts, and Cataclysms

Scope:

Impact craters exist on all the solid bodies of the solar system, from asteroids to moons to planets. Their ubiquitous nature and ancient history indicate an epoch of continuous bombardment during the formation of our planetary system. Such impacts continue to this day with reduced frequency but sometimes devastating consequences. We have witnessed one such massive collision on Jupiter, and we find geologic evidence for such events in Earth's past. The global reach of a major impact on Earth can include fires, earthquakes, tidal waves, and global dust clouds. The combined effect can produce a mass extinction that would wipe out nearly all the species on our planet. Future impacts are not a question of if they will happen, but rather when they will happen.

Outline

I. Impact craters are the most abundant surface feature in the solar system.
   A. Hundreds of impact craters have been found on Earth.
      1. The rule of thumb is that the crater is about 20 times larger than the impacting body.
      2. Easily recognizable craters are found for the most recent impacts (from tens of thousands of years ago).
      3. Older impact structures can be recognized from aerial and satellite imagery, though erosion can make them hard to spot.
   B. The bowl shape of a crater derives from a high-energy impact.
      1. The speed of impact is comparable to the Earth's speed through the solar system: about 100,000 kilometers per hour.
      2. Generally, several megatons (or more) of energy are released in the impact.
      3. The material excavated from the crater is strewn across the surrounding region.

- **C.** Craters are a significant feature on all of the rocky planets.
    1. Venus has similar numbers of craters as Earth, but those on Venus are better preserved.
    2. Craters on Mars are heavily concentrated in the southern hemisphere—which features Hellas Basin, one of the largest impact structures known.
    3. Mercury and the Moon appear similar, as both are heavily cratered over most of their surface.
    4. The near side of the Moon shows many large impacts that filled in with lava flows, while the far side shows only a few.
- **D.** Many moons and asteroids show abundant cratering.

II. The historical record of cratering can teach us the history of the solar system, planets, and the evolution of life.
- **A.** Dating the craters on the Moon shows an epoch of intense cratering following the formation of the solar system.
    1. The heavy bombardment era lasted until about 3.5 billion years ago, and the cratering since that time has been relatively constant.
    2. The fact that cratering has not slowed down means that new Earth-crossing asteroids and comets must be continually created.
- **B.** An explosion in Tunguska in 1908 was the most powerful recent impact on Earth.
    1. A roughly 50-meter-sized impactor exploded about 8 kilometers above Earth's surface.
    2. The explosion was about 1000 times more powerful than Hiroshima and Nagasaki combined.
- **C.** In 1994, we were lucky enough to witness an impact event on Jupiter.
    1. Comet Shoemaker-Levy 9 was captured by Jupiter's gravity and broke up while in orbit.
    2. A collection of 2-kilometer sized fragments of a comet crashed into Jupiter, creating Earth-sized dark spots in Jupiter's atmosphere that dissipated over a few months.
- **D.** On Earth, an impact event is the probable explanation for the end of the age of dinosaurs.

1. Tektites are beads of once-molten mineral glass that solidified in midair after an impact. Finding them is evidence of a nearby impact structure.
2. The Chicxulub impact structure, on the Yucatán Peninsula of Mexico, is about 200 kilometers across.
3. Chicxulub dates to 65 million years ago, the time of the mass extinction at the end of the Cretaceous period.
4. The geologic record shows a worldwide layer of iridium and other rare elements deposited at the KT boundary.
5. Some volcanologists disagree with the impact hypothesis. They argue that the gases released by extensive volcanoes in India changed the climate and caused the Cretaceous-Tertiary mass extinction.

E. Impact events appear to be associated with other mass extinctions on Earth.
1. Metal enrichments occur in the strata associated with mass extinctions 11, 35, 65, and 91 million years ago.
2. Three layers of impact melt are found in sediments associated with the mass extinction 35 million years ago.

III. The effects of asteroid and comet impacts on Earth can be global.
A. It is interesting to contemplate how a 10-kilometer-sized impactor can have global effects on a 13,000-kilometer-sized planet.
B. The energy released in such a large impact is equivalent to 100 million nuclear bombs.
1. The impact excavates a crater up to several hundred kilometers across.
2. The ejecta from the crater rain down over a wide region.
C. The speed of the impact creates a fireball while streaking through the atmosphere and a blast wave upon impact.
1. The searing blast wave will instantly incinerate the nearby landscape.
2. Interaction with nitrogen in the atmosphere and sulfur in the ground can lead to widespread acid rains.
3. Earthquakes from a large impact could be about 100 times more powerful than the largest ever recorded.

- **D.** Two-thirds of all impacts would be in the oceans.
    1. Tidal waves kicked up would be kilometers in height and sweep across continents.
    2. Impacts in the oceans can throw enough water into the atmosphere to create a temporary increase of greenhouse warming of up to 10°C (18°F).
- **E.** Atmospheric effects are difficult to calculate but can be particularly deadly.
    1. Dust clouds kicked up by the impact can circulate globally.
    2. Fires ignited by the blast wave and friction-heated ejecta would exacerbate the problem.
- **F.** Disruption of the climate and the food chain can lead to global mass extinctions.

IV. Future impacts are only a matter of time.
- **A.** Meteors are tiny impacts that burn up in the atmosphere and occur daily.
- **B.** A Tunguska-level event should occur every few hundred years.
- **C.** An impact large enough to wipe out a city should occur every 10,000 years.
- **D.** Country-wide or regional destruction events are estimated at every million years.
- **E.** Impacts with global consequences are predicted to occur about once every 100 million years.
- **F.** Currently, no large asteroids are coming close. The previously sparse monitoring programs are improving steadily.
- **G.** Given sufficient advance warning, we could nudge the potential impactor into a safe orbit.

Suggested Readings:

Beatty, Petersen, and Chaikin, *The New Solar System*, chap. 6.

Benson, *Beyond: Visions of Interplanetary Probes.*

Hey, *Solar System.*

Lewis, *Rain of Iron and Ice.*

Light, *Full Moon.*

McFadden, Weissman, and Johnson, *Encyclopedia of the Solar System*, chap. 43.

Near-Earth Asteroid Tracking. http://neat.jpl.nasa.gov/

Near Earth Object Program. http://neo.jpl.nasa.gov/

Verschuur, *Impact!*

Questions to Consider:

1. How would Earth's surface look if there were no erosion to erase craters?

2. Some craters on moons are a quarter (or more) of the size of the moon. How do you think large impacts have affected the number and sizes of moons?

3. Given that impacts are a long-term consideration with low probability of immediate consequences, how large of a budget should we devote to monitoring the threat of asteroid or comet impact?

Lecture Ten—Transcript
Craters, Impacts, and Cataclysms

The process of science is a relatively calm endeavor. It involves much quiet thinking and contemplation. Personally, I do some of my best work in the early morning, when the world around is still and I can focus intently on the ideas at hand. Sometimes by the time my wife and kids get up, the most productive part of my day is already over. I know one astronomer who will not get cable TV. It's not because he doesn't like television. It's actually because he thinks he will like it, and he doesn't want to have it there as a distraction, so that he can focus on doing his science.

It's also true that the results of science rarely have immediate consequences. Most astronomical objects that we observe will change very little on timescales of millions or even billions of years, much less on human timescales. It's not very likely that you'll hear somebody say, "We have an emergency. Quick, call an astrophysicist." That's not something I really expect to hear in my lifetime, but there is one topic where Hollywood has latched onto science, where discoveries can be of immediate and wide importance, and where astronomy is considered exciting enough for an action movie.

That is the topic of an asteroid impact on Earth. Nothing like the threat of death, destruction, and devastation to garner a little attention. Hollywood, as usual, gets the science details mostly wrong, or they enhance them to make a good story. I can't really blame them because their job is to sell you entertainment. But, as a scientist, I would like to sort of correct the misperceptions that Hollywood has propagated.

In this lecture, I'd like to review the real story. I'd like to show you the effects of impacts across the solar system and discuss the scientific view of the threat posed by asteroids and comets. Impact craters are the most abundant surface feature across the solar system. On Earth, we've identified hundreds of impact craters.

Here is one. It is Barringer Crater in northwest Arizona. It's about 1.2 kilometers across and 170 meters deep. If we look from the rim of the crater, what do you see? It's a great big hole in the ground. However, this was the first structure shown to be of impact origin, and it's also the best-preserved impact structure in the world. It was caused by the impact of a nickel-iron meteorite about 50,000 years ago.

The impactor is estimated to be about 50 meters across. Now, we have no real evidence of the actual size of the impactor. That just comes from a rule of thumb—the crater is about 20 times larger than the impacting object. We can't tell you what the impacting object looked like because not much remains of it. Most of it was vaporized upon impact. When they first discovered Barringer Crater, there were about 30 tons of scattered remains on the plains around it. It sounds like a lot, but it's really tiny compared to the 300,000 tons that were originally contained in the impactor.

The crater was originally thought to be of volcanic origin, and its impact origin was not suggested until 1906. It took another five decades for the impact origin to be proven, and that was proven not because of the crater itself but because they found shocked quartz crystals. These are quartz crystals that heated up to a really high temperature and indicated a very high-energy event. And you can't get those from volcanoes. You can only get them from impacts.

The impact process creates the standard bowl shape of a crater. Here is a diagram of this impact process. The speed of the impact is going to be comparable to the speed of Earth's orbit around the Sun. The Earth is moving at about 30 kilometers per second in its orbit around the Sun, so another object that we strike is going to be moving at approximately that same relative speed.

This is about 100,000 kilometers per hour. The energy released when it strikes the surface is going to be several megatons. That's equivalent to the explosion of a nuclear bomb. The shock wave will press down and out through the surface and will carve out the bowl shape. The material that is then pushed will spray out the sides to become ejecta that will land on the region around it. The rebound from the shock wave can come back, and for the larger impacts, it can actually form a central peak in the center of the crater. The steep sides of the bowl will often collapse over time.

If we look at Barringer Crater from satellite photos, it looks like this. And you can definitely see the nice circular feature. But only the recent impacts can be easily recognizable. These are impacts within the last 100,000 years. The erosion due to wind, rain, snow, and other forces makes the older ones hard to spot.

Here is one in the outback of Australia. Now, the outback is old terrain, so it's a good place to search for very old things. This is Gosses Bluff Impact Crater, and it's dated about 142 million years ago. The crater itself is about 20 kilometers across, and that central ring is easily visible. But you can see that over time, the outer circular features are more subtle. If you look carefully, you can see that there are circular features around those central circular features.

Even older impact structures can really become part of the landscape. Here is an impact crater called Manicouagan Lake in Canada, and it is this lake feature here, this circular lake. This was formed about 214 million years ago, and the original crater was probably about 100 kilometers across. It has since collapsed to be about 70 kilometers across. Within the central island, there is a peak called Mount Babel, and we believe this to be the central peak of the impact crater. Now, the lake makes this crater much easier to recognize. It may not have been so easily recognized before the crater was flooded in order to create a hydroelectric project.

Some other craters are impossible to recognize from the ground. This image here is one of a crater called Kebira. Now, there was this mystery of what they called Libyan desert glass, these wonderful, beautiful glass beads, but they were of unknown origin. Now, somebody who knew something about impacts saw them and asked a geologist who was working in the area: Oh, is there an impact structure around here? And the geologist said: No, no, no, there's no impact structure.

But that thought didn't leave him, and he went searching through satellite photographs until he found this crater. It is 31 kilometers across, and he believes that it was impossible to recognize from the ground because it was just so large. And so he named it Kebira, which means "large" in Arabic. It is the largest crater in the Sahara, and it's estimated to be tens of millions of years old. You can see that where the impactor hits decides whether or not it's more or less easily recognizable.

You can also note that since Earth's surface is covered two-thirds with water, most of the impact structures on Earth are not going to be visible. Now, beyond the erosional processes, the impacts on Earth are also limited by its atmosphere. Only objects larger than a few tens of meters can actually survive passage through the atmosphere, and those that do can lose half or more of their mass. The 50-meter impactor that hit at Barringer Crater was probably a significantly larger asteroid when in space.

Now, craters are a significant feature on all solid surfaces across the solar system. On Venus, they're especially well preserved. Here is a picture of three impact craters on Venus, and Venus has similar numbers of craters as Earth, but as I said, they're better preserved because there is very little erosion. Venus has a very thick atmosphere, but at its surface the winds actually blow very gently. There's not much to erode away these craters. Matter of fact, the most important erosion process on Venus is lava flows, and those seem to happen infrequently.

We also see impact craters on Mars, and here's a very beautiful one on Mars. It's the standard impact structure. It's called Victoria Crater, and it's about 800 meters across. But this one shows a very unusual feature. If we zoom in to the northern rim, we see these strange markings along it, very linear features. Are these evidence of Martians? No, unless the Martians happen to be made in the United States. These are the tracks of the Opportunity rover exploring the rim of Victoria Crater, as seen by the orbiting satellite around Mars. Here is a view of the Victoria Crater from the ground, from the Opportunity rover.

Mars also features some very large craters. This one is called Hellas Basin, and it's one of the largest impact structures known. It's about 2300 kilometers across, and it's sort of a little hard to see in a visible-light image. You can see—oh yes, there's this sort of roundish feature here. But if I show you a Mars topographic map, it stands out very strongly, and it's this giant blue region down here. It is about 9 kilometers deep—and at 2300 kilometers across, the volume of this crater is about 200 million times larger than the volume of Barringer Crater.

You can see other large impact craters on Mars. There's one here, just to the left of center, and there is a region up here, just above the Hellas Basin, that looks like that could also have been caused by a large impact crater. But what's most striking when you look at this topographic map of Mars is how the top of it is incredibly smooth—these are smooth lowland volcanic plains—and the bottom of it is incredibly cratered, the heavily cratered highlands.

This is a little bit of a puzzle to me, and it has sort of bothered me every time I see it. If you want lots and lots of cratering, you go take a look at Mercury and the Moon. On Mercury, there's a very famous large impact structure called Caloris Basin. It's kind of hard to see in this image because it straddles the terminator, and if you can see a circular feature—let me draw a circle over it to guide your eye—this is about 1300 kilometers across, and it's about one-quarter the size of the entire planet. The giant impact structure is one-quarter the size of Mercury's diameter.

Mercury is heavily cratered across its entire surface. Here is an image of a crater field, and you can see just craters upon craters upon craters. Most people when they see Mercury actually think it's the Moon because the Moon's surface looks very similar. So, for comparison, here is a crater field on the Moon, and again, you see craters upon craters upon craters. The interesting thing that you see on the Moon is that the interiors of some of the craters seem smoother than they seem on Mercury. There have been more lava flows on the surface of the Moon than there have been on the surface of Mercury.

What this tells you is that there's a long history of impacts recorded on the surface of the Moon, and we'll get back to that in a few minutes. Looking at the Moon globally, we can look at its near side, and we can see many large impacts. These large, smooth gray areas are very large impacts that have filled in with lava. These are the lunar seas that they call maria. They, of course, aren't seas. They're just lava flows that have filled in.

In contrast, when we look at the far side, we see only a couple large impacts that have lava flows filled in. So there is a dichotomy between the side of the Moon we always see and the side of the Moon that we didn't get to see until the Apollo era. Until the space age, we had never seen the back side of the Moon. So why is this true? Well, there are two reasons. One is that on the near side of the Moon, the crust is a little bit thinner. If you actually measure the thickness of the crust near side versus far side, the crust is a little bit thinner, so large impacts can punch through the crust easier on the near side.

The other reason has to do with the fact that the Moon is tidally locked to Earth. The Moon always shows the same face to Earth, and objects passing by Earth can be gravitationally focused toward the near side of the Moon. So, we would send more impactors towards the near side of the Moon, and because they swing past Earth, they would get a gravitational boost, and they would arrive at faster speeds. Both of these would lead to more large impacts on the near side of the Moon.

Now, the Moon has one other interesting feature that I'd like to point out to you. It's called Mare Orientale, and it's one of the classic large impact basins, but it's very hard for us to see here from Earth because it's basically on the eastern edge of the Moon. Here it is in the Clementine data, and you can see it's a large, ringed impact structure about 900 kilometers across. The impact sent ripples across the lunar surface and created these several rings here.

This reminds me of another large impact structure on Jupiter's moon Callisto—here is probably the most impressive ringed impact structure in the solar system. It's called Valhalla on Callisto. Callisto is Jupiter's outermost of the four Galilean moons. It has a very icy surface. Now, the bright spot in the center is about 600 kilometers across, but you can see ring fractures that have spread out to about 1500 kilometers in radius. That's about 3000 kilometers across for the effects of this impact structure on this small moon.

Even small bodies of the solar system show craters. This is an asteroid named Gaspra. The asteroid itself is only 20 kilometers across, much smaller than some of the impact structures I've shown you. However, it itself is still pockmarked by even smaller craters, so even the impactors have suffered impacts before they do their job of creating a crater.

Now, let's take a look at the historical record of cratering in the solar system. The place to go, the place to look, is the Moon. The Moon provides a rich history of cratering, where we can age date the craters, simply because the younger craters will be on top of the older craters, and so we can then piece them together to assemble the number of craters over time. The results show that there's been a steady component to the cratering rate as you go back in time. Back to about 3.5 billion years ago, there's been a continual cratering, not really much change in the rate.

Before that time, there's an exponential increase in the number of craters we see on the Moon. This indicates a period of late heavy bombardment. This was about 3.5 billion years ago that there was an intense cratering period in our solar system. The Moon formed about 4.5 billion years ago, so for about a billion years, the solar system was sweeping up its debris, and there were incredible amounts of impacts slamming all the bodies of the solar system. Basically Earth, Moon, Venus—all the major bodies should have been continually pounded until the late heavy bombardment came to an end.

Now, the steady cratering shows another thing. It shows that there must be population renewal. The objects that are hitting us today have to be renewed, otherwise they would die out. If there was only a small population hitting, after a while, they'd be gone. So we have to renew that population. That's an important thing that we'll come to in a later lecture.

Well, that's the long-term history. What about recent history in terms of cratering? Well, in 1908, on June 30, there was a massive explosion over the skies of a remote part of Siberia. Booms were heard by many, a shock wave knocked people down, and earthquakes were registered at many seismic stations. In 1927, they made a scientific expedition to the area and found this: that 19 years later, there were still felled trees from that explosion. From the pattern of trees and from other evidence, they deduced that a roughly 50-meter-sized impactor exploded about 8 kilometers above Earth's surface.

There's no obvious impact crater, although some scientists recently have suggested a certain lake nearby the epicenter of the explosion, to be where what was left of the impactor landed. It hasn't been proven yet. But this is the largest recent impact event on Earth, and it was about 1000 times more powerful than the bombs at Hiroshima and Nagasaki combined. Such a small impactor would be very difficult to find and track, so it could come without warning, and it could destroy a city.

This is the kind of event that inspires Hollywood. There, of course, have been other events, and in 1994 we were lucky enough to witness an impact event on Jupiter. This is a montage; it's not a single image. It's a montage of an image of Jupiter and an image of Comet Shoemaker-Levy 9. Shoemaker-Levy 9 was captured by Jupiter's gravity and broke up while in orbit. Now, we did not discover S-L 9 until after it had broken up. And once we discovered it, we were able to predict that it would collide with Jupiter 16 months later.

Now, think about this. What's going to happen? We've got 2-kilometer-sized fragments of a comet, and then we've got Jupiter's massive atmosphere. Is it going to be exciting? Is it going to be a huge impact? Well, honestly, at the time we were looking at this, we weren't really sure whether it was going to be an event or a nonevent. That didn't stop us. We turned every telescope possible to watch it.

And with the Hubble Space Telescope, we were able to see this. In the center, you can see the disc of Jupiter, and along the lower portion of Jupiter, you see some brown spots. These brown spots are the impact sites where S-L 9's fragments slammed into Jupiter. It's much more visible on the right-hand side, where you see Jupiter in the ultraviolet. And here you can see the giant holes in Jupiter's cloud tops that were created by the impacts.

For scale, here is Earth to the correct size. These impact events scarred Jupiter, but they dissipated over the course of a few months. It was really the same thing as happened in Tunguska. They had giant explosions in Jupiter's atmosphere that caused these big scars. Eventually, the clouds reformed, and Jupiter went on as if nothing had happened—but it was really exciting at the time, I have got to tell you. It's one of the few times in astronomy where it was like events happening in real time.

The final historical event I'd like to discuss is one where sometimes you have evidence of an impact, but you don't know where the crater is. Now, tektites are a strong clue of an impact. They are beads of mineral glass that were once molten and had to solidify in midair. The glass was hot enough as it was flying through the air it then solidified. For example, there's a tektite field across North America, and it was recently traced to a crater in the Chesapeake Bay.

Tektites had also been found on the island of Haiti, and we assumed that the crater was somewhere underwater. We were half right. Here is a relief map of the Yucatán Peninsula. If you look closely, you can see somewhat of a circular feature. Let me again guide your eye, and you can see that there's a circle here. Half of the crater is underwater. It becomes much easier to see if instead we look at a gravity map.

This is how we first discovered this crater. This was not done by research scientists. It was actually done by folks working for petroleum companies. They were looking for oil off the Yucatán

Peninsula, and they found this impact structure. Nowadays, with computer processing, we can actually create a 3-D map. Here is a 3-D map of what we call the Chicxulub impact crater, and you get a multiringed impact structure about 200 kilometers across. This is nearly three times the diameter of Lake Manicouagan. This is a truly huge impact event, and it has been age dated to 65 million years ago, the time of the Cretaceous-Tertiary extinction.

This was the time of the end of the age of the dinosaurs. The question was: Could this impact have been the downfall of the dinosaurs? The key piece of evidence in that argument is a layer of iridium found in the KT boundary layer. Basically, as the sediments are deposited, when you look through the sediments at the KT boundary layer, from the Cretaceous to the Tertiary, you find a thin layer of iridium. This has been found at over 80 sites worldwide.

Now, iridium is extremely rare on Earth, but it is common in iron meteorites. The excess was calculated to be what would correspond to about a 10-kilometer asteroid. And a 10-kilometer asteroid would make a 200-kilometer crater. So it appears that while the dinosaurs ruled the planet for 165 million years, their doom may have come suddenly and swiftly with an impact event.

I must note, however, that the volcanologists out there will disagree. They would point to the Deccan Traps lava field near Mumbai in India as the source of the extinction. These lava fields cover hundreds of kilometers, and they are also age dated to the time of the KT extinction. They would argue that the carbon dioxide and the sulfur dioxide released in these volcanic eruptions changed the climate and caused the extinction.

Now, I'm not going to get involved in this because I'm not expert enough to evaluate the details of one argument versus another. I will say that as an astronomer, I'm naturally biased toward impacts. I can certainly see that perhaps there were multiple factors contributing to this worldwide mass extinction.

Furthermore, my astronomer friends have told me that impacts appear associated with other extinctions. They find metal enrichments that occur in the strata associated with extinctions at 11, 35, 65, of course, and 91 million years ago. Of importance, there are three layers of impact melt associated with a 35-million-years-ago extinction. So, based on that combination of evidence, it does appear that the effects of asteroid and comet impacts on Earth can be global. But wait—you shouldn't just accept that. You've got to think about that.

We have a 10-kilometer-sized impactor. We have a 13,000-kilometer-sized planet. I mean, how is that going to create a global effect? I mean it's like a pinprick. How? Well, here's how. The energy released in such a large impact is equivalent to 100 million nuclear bombs. The crater is about 20 times the size of the impacting object. Now, this spreads stuff out over hundreds of kilometers, and it may be a relatively localized effect, but the ejecta strewn out from within that crater will rain down over a wide region. That'll cover thousands of kilometers with hundreds of millions of tons of material.

The blast wave created will incinerate the nearby landscape. Plus the fireball, as it's streaking through the atmosphere, will interact with the nitrogen in the air and the sulfur in the ground and create widespread acid rain. As the impact hits the ground, you create an earthquake. They estimate that the KT impact, the earthquake, could have been as large as magnitude 11 or 12. The largest earthquake ever recorded is 9.5, and the earthquake scale is logarithmic, so the earthquake could have been 100 times more powerful than the largest ever recorded.

Furthermore, if it struck in a shallow sea, as we believe it might have, it would kick up a tsunami. Tsunamis kicked up by an impact of this magnitude would be a kilometer in height and would be able to sweep across continents. Furthermore, when you hit in the ocean, you kick up a huge amount of water into the atmosphere. That water is an incredible greenhouse effect. I know we talk about carbon dioxide, but water is an even better greenhouse gas than carbon dioxide, and that can provide temporary greenhouse warming of 10°C, or 18°F.

Furthermore, there would be fires. The blast wave would ignite some fires, but also the ejecta would be streaking through the atmosphere, would be heated to incredible temperatures, and would become red hot. When it lands, it would instantly ignite the vegetation around it. Then there would be dust clouds, first of all kicked up by the impact—and the smoke from the fires. Through the global circulation, that would spread all around the globe and block out the Sun.

So, what really happened at the KT impact was that you disrupted the food chain because there's evidence that the flowering plants disappeared for thousands of years. If the plants die, then the plant eaters die. If the herbivores die, then the carnivores die. And that leads to mass extinction.

So, these are the horsemen of the asteroidal apocalypse: craters, ejecta, blast waves, earthquakes, tsunamis, fires, dust clouds. And their combined effect is how a 10-kilometer-sized body can have global consequences on a 13,000-kilometer-sized planet.

Now, naturally, your next question is going to be: When will this happen again? And future impacts are really only a matter of time. We have impacts every day. They're called meteors, and they streak through the atmosphere and burn up. A Tunguska-level event should occur every few hundred years. An impact large enough to wipe out a whole city should occur every 10,000 years. If we want to think even bigger, the country-wide or a regional destruction event would happen roughly every million years, but the impacts with global consequences are not so common. They should occur once every 100 million years.

The last one was about 65 million years ago, but the occurrence of the next one is probabilistic. It could be 50 years from now, or it could be 50 million years from now. We do know that currently there are no large asteroids coming close to our planet. And given funding, we can monitor the large objects easily.

I was surprised when I checked the near-Earth asteroid website, and the technology has developed faster than I expected, and some of the objects we're tracking are as a small as a few meters across. Now, of course, these require continuous monitoring, as the asteroids' orbits tend to shift a bit, but we may be able to see the next Tunguska-level event before it hits us. And with advance warning, we might be able to send a spaceship up and sort of nudge it out of its orbit so that it would move into a harmless orbit.

We might be able to prevent the next major asteroid hit, so I guess the moral from this is that the astronomers and NASA need your support. After all, the dinosaurs seem to have died out because they didn't have a space program to prevent it.

All right. Next time, we'll take a look at the planetary interiors. We'll see you then. Take care.

Lecture Eleven
Journey to the Centers of Planets

Scope:

The composition, topography, and motions on Earth's surface indicate a dynamic interaction with the interior that is encapsulated in the ideas of plate tectonics. For Earth, and the rest of the rocky planets, the surface we see is a thin crust layer whose depth is roughly 1% of the radius of the planet. Nearly all the volume is in a rocky mantle and an iron core. This differentiated structure is replicated for the larger moons and perhaps the large asteroids, but the smaller and lower-mass bodies are expected to be roughly the same mixture of materials throughout. The giant planets, while gaseous on the exterior, are actually more liquid in the interior. Only at the very core is a chunk of rock comparable to several or many Earths. Charged particles in the interiors of some planets, coupled with the planet's rotation, create planetary-scale magnetic fields extending from the interior to well beyond the surface.

Outline

I. Earth's surface provides many clues about its interior.
   A. The three main types of rocks imply a cycling of material between surface and interior.
      1. Igneous rocks are those that cooled from molten lava or magma.
      2. Sedimentary rocks are deposited by wind, water, and/or snow (or ice) over time.
      3. Metamorphic rocks are igneous or sedimentary rocks that have been subjected to high pressure and temperature and changed their structure.
      4. The implications from these rock types is that magma must flow up from the interior and that surface rocks must flow down into the interior.
   B. The topography of the ocean floor reveals a system of interconnected features.
      1. The deepest parts of the oceans are trenches that generally occur along the edges of continents.

- 2. From the deepest ocean trenches to the highest mountains is about 20 kilometers of vertical relief, but that is still only about 0.3% of Earth's radius.
- 3. Mid-ocean ridges are like undersea mountain ranges rising a kilometer or two above the ocean floor.

C. We interpret the ocean's topographic features as boundaries of large plates.
- 1. Earth's crust is divided into 8 large plates and a few smaller ones.
- 2. Trenches are regions where one plate is subducting underneath another plate.
- 3. Ridges are regions where magma rises from the interior to create new plate material.
- 4. The number of and strength of earthquakes are highest at the plate boundaries.
- 5. Volcanic activity is associated with trenches.
- 6. The long chain of undersea mountains leading to Hawaii is an example of hot-spot volcanism that tracks the motion of the Pacific Plate over many millions of years.

D. The history of plate motions shows that Earth's continents are floating atop a fluid interior layer.
- 1. The idea of continental drift is easily imagined by noting that the bulge of the east coast of South America would fit neatly into the west coast of Africa.
- 2. About 180 million years ago, all the continents were together in one supercontinent called Pangea.
- 3. Where a continent is today has little to no correlation with where it was a few hundred million years ago.

II. Our detailed knowledge of Earth's interior serves as the template for understanding other worlds.

A. Differentiation is the gravitational separation of materials of differing densities.
- 1. For differentiation to occur, the body must have been molten at some point during its formation.
- 2. We expect planets and larger moons to be differentiated, while the smaller moons, asteroids, and comets should be undifferentiated.

- **B.** Seismology measurements probe Earth's interior.
  1. Earthquakes trigger seismic waves that pass along Earth's surface and through its interior.
  2. The properties of seismic waves, notably their speed of propagation, depend on the medium through which they are traveling.
- **C.** The interior structure of Earth has three main layers.
  1. The crust is roughly 60 kilometers thick and is about 1% of Earth's radius.
  2. The crust is composed of low-density materials and contains only 0.4% of Earth's mass.
  3. The rocky mantle extends about halfway down and contains about two-thirds of Earth's mass.
  4. The upper mantle contains the asthenosphere, a region that acts like a fluid on geologic timescales.
  5. The core is composed of iron and nickel and holds about 30% of Earth's mass.

III. The interiors of other large rocky and icy worlds are comparable to Earth's interior structure.
- **A.** The other rocky planets are all differentiated, but each shows distinct variations in interior structure and motions.
  1. Venus's interior structure is thought to be similar to Earth's, with the major difference being that it is a one-plate planet.
  2. Surface features on Venus are likely due to convective motions in the mantle and episodic volcanic resurfacing.
  3. Mercury's small size and high density point to a large iron core extending more than two-thirds of the planet's radius.
  4. Long faults, called scarps, on Mercury indicate compression of its crust during solidification.
  5. Mars has a lower density that indicates a lower percentage of iron in its composition.
  6. The uplifted and volcanic regions on Mars are confined to two major areas that likely indicate sustained regional motions in the mantle.
  7. Compared to Earth's plate tectonics, the other rocky planets' surfaces show relatively mild expressions of their interiors.

B. The large moons and minor planets illustrate a sequence from differentiated to undifferentiated bodies.
1. The Moon is differentiated, but its lower density means it probably does not have a core.
2. The Moon shows evidence of ancient crustal extension and contraction but no major internally generated geologic activity for the past 2.5 billion years.
3. The largest asteroids are likely to be differentiated. The existence of iron meteorites provides evidence of the complete breakup of some differentiated bodies.
4. With some exceptions, the Galilean moons of Jupiter are thought to have iron cores surrounded by rock, with icy surfaces.
5. Europa, Ganymede, and Callisto show potential evidence for having subsurface oceans.
6. Titan is believed to have a differentiated rock and ice interior with a subsurface ocean of water and ammonia.
7. Other large bodies of the outer solar system, like Triton and Pluto, have densities indicative of undifferentiated ice and rock.

IV. The giant planets (sometimes called "gas giants") are actually more liquid in their interior.
A. Like the Sun, the giant planets have an interior structure determined by hydrostatic equilibrium.
B. Jupiter and Saturn have roughly the solar composition of hydrogen and helium.
1. The outermost layer is gaseous molecular hydrogen.
2. At higher pressures, hydrogen breaks down into a liquid called metallic hydrogen.
3. Jupiter is predominantly liquid hydrogen (by volume), whereas Saturn is more gaseous hydrogen (by volume).
C. Uranus and Neptune are significantly smaller and resemble the cores of Jupiter and Saturn.
1. They probably have a small rocky core of a few Earth masses. They form farther out in the solar system, and instead of accreting gas, because they didn't have enough mass, they instead accreted "ices."
2. The dominant component is a hot, highly compressed liquid of "ices": a mixture of water, ammonia, and

methane, along with various chemical species derived from these.
3. The outer 15% of these planets is molecular hydrogen gas.

V. Magnetic fields are generated in the interior of most planets. They are an external manifestation of interior structure, and one which, as we'll see in a later lecture, provides protection from interplanetary storms.

Suggested Readings:

Beatty, Petersen, and Chaikin, *The New Solar System*, chaps. 12, 14.

Bennett, Donahue, Schneider, and Voit, *The Cosmic Perspective*, chap. 9.

De Pater and Lissauer, *Planetary Sciences*, chap. 6.

Hey, *Solar System*.

McFadden, Weissman, and Johnson, *Encyclopedia of the Solar System*, chaps. 8, 10, 16, 21.

Shu, *The Physical Universe*, chap. 17.

Questions to Consider:

1. How does the composition of an object, from rocks to ices to gases, affect its formation and its interior structure?
2. What role might the impacts of the heavy bombardment era play in creating differentiated worlds?
3. Anything falling into a giant planet would be crushed by high pressure well before it hit any sort of solid surface. Would the giant planets qualify as a practical version of a bottomless pit?

Lecture Eleven—Transcript
Journey to the Centers of Planets

In 1864, Jules Verne published the book *A Journey to the Center of the Earth*. In it, Professor Lidenbrock leads his nephew and a hired guide down a volcano in Iceland, and they have many adventures, where they sail across a subterranean ocean, they meet prehistoric creatures, they are hit by underground lightning, and they see a 12-foot-tall ape-man, which they can't decide if it's an ape-man or a man-ape. And upon looking at that, I wondered: Is that related to the fact that Darwin's *On the Origin of Species* was published five years earlier? But at the end of their adventures, they come to the surface in southern Italy by a steam vent of a volcano.

Now, unfortunately, I don't think truth is stranger than fiction in this case because as we explore planetary interiors we will see a subterranean ocean, but it will be made of magma, not water, so you couldn't really sail across it. Instead of lightning, we'll find magnetism. And we won't meet any prehistoric creatures, although there is a chance of life on Europa.

Our trip to the interior will be based on bulk measures and external properties because really, we have no chance of making in situ measurements in the interior of any planet. We'll spend most of our time looking at Earth, where the data is strongest, and as we extrapolate to the other planets and bodies of the solar system, we'll have to be guided by mathematical models. And generally, when we look at the interior, we're really biased towards explaining the surface characteristics of the planets.

So, we'll begin where Verne begins his adventure, on Earth's surface. Earth is a rocky planet, so let's take a look at the rocks and see what we can tell about the interior from just looking at the rocks. There are three main types of rocks: igneous, sedimentary, and metamorphic. Igneous rocks are rocks that cooled from a molten state. Now, molten rock is called lava when it flows across the surface, and it's called magma when it's beneath the surface. Examples of igneous rocks include basalt and granite.

Sedimentary rocks are rocks that have been deposited over time by wind, water, snow. For example, layers of sand can be compressed, and you get sandstone. If the minerals precipitate out of the oceans and form on the ocean floor, you create limestone. It's interesting to note that the rocks at the top of Mount Everest are limestone. These are sedimentary rocks that formed at the bottom of an ocean and then were pushed up to form the highest mountain in the world.

Metamorphic rocks are reprocessed rocks. If you subject igneous or sedimentary rocks to high pressures and temperatures, they will change their structure. Fine-grained igneous rocks become a different type of rock called schist. Limestone becomes marble—so when you're picking out your countertops, and you have a choice between granite and marble countertops, here's the difference. Granite are igneous rocks. They've been molten once, and then they solidified, while marble is a metamorphic rock that was processed as a sedimentary rock, and then reprocessed to form marble. Probably that extra step in production means that marble costs a little bit extra.

The implications from rocks for the interior of our planet are twofold. One, it shows that magma flows up from the interior. There must be some sort of magma ocean beneath our water oceans, where magma can rise up and form our igneous rocks. The second thing is that rocks must also flow down into the interior. Some of the crust is recycled because you're taking these igneous and sedimentary rocks and pushing them down into high temperatures and pressure to become the metamorphic rocks.

Next, let's take a look at the oceans. Here is the standard picture of Earth's surface, and you can see the oceans cover 70% of the surface. There are five major ones: the Pacific, the Atlantic, the Indian, and down near Antarctica there is the Southern Ocean that we don't talk about very much, and then at the northern latitudes we have the Arctic Ocean.

The average depths of these oceans are about 13,000 to 16,000 feet, about the height of a really good mountain, but this is actually depth in the sea. The only one that's different is the Arctic Ocean, and it's a rather shallow ocean. It goes only to about 4000 feet deep. The deepest points of the ocean are shown in this topo. map, where you can see trenches.

Now, trenches tend to occur along the edges of continents. Along the west coast of South America, you can see the Peru-Chile Trench. Along the west coast of North America and stretching along past Alaska is the Aleutian Trench. And then down past Japan, you have the Japanese Trench, and stretching out into the Pacific, you have the

Mariana Trench. The depths of these trenches can reach to 36,000 feet, and that sounds really deep, but let's put it in perspective with respect to our planet.

The average depth of the ocean is perhaps 5 kilometers, and the deepest is only 11 kilometers. Now, let's add in the mountains: 11 kilometers deep for the oceans, but you know, add in the 9 kilometers of height for the mountains, and you get about 20 kilometers of vertical relief. That's really only 0.3% of Earth's 64-kilometer radius.

Now, I have here a globe, and some globes have vertical relief on them, where you can feel the mountain chains. If this globe had perfectly accurate vertical relief, going from the highest mountains to the deepest trenches, the thickness on a 12-inch globe, or 6-inch radius, would be less than half a millimeter. You could rub your hand across it and you still could not feel it. That's why most globes have really, really, really exaggerated vertical relief.

The other features of the seafloor are what we call ridges. In this topo. map, we emphasize the ridges, and you see, running down the center of the Atlantic Ocean, the Mid-Atlantic Ridge. And off the west coast of South America, we have the East Pacific Rise. And flowing underneath Australia and India, we have the ridges in the East Indian Ocean. These are like undersea mountain ranges. They can reach up to 5000 feet above the seafloor. And sometimes, especially in the Mid-Atlantic Ridge, they're cut by a rift valley. So you have the mountains, and then you have a rift valley cutting through it.

We interpret these ocean features under the guise of plate tectonics. In this image, we show you the plates of Earth. Earth's crust is divided into eight large plates and a few smaller ones. We have the North American Plate and the South American Plate. We have the Eurasian Plate, which stretches from Europe across Asia, and the African Plate. We have the Indo-Australian Plate, from India and Australia, and we have the Pacific Plate down on the bottom. And then we have one plate that's just to the west of South America, which is called the Nazca Plate. Those are the eight major plates.

There are also a host of minor plates, and the most easily recognizable one is the Arabian Plate that includes Saudi Arabia. Now, with these plates, you can see the boundaries of these plates fit with the ridges, and where these plates intersect are the regions of the trenches. The trenches are interpreted as regions of subduction, where one plate is going underneath another.

The ridges are the creation of new plate material. The magma rises from the interior and pushes up these mountains. It solidifies and creates new plate material, and then the plates spread away from the ridges. This motion can be measured, and it's about a few centimeters per year. So if you think your flight from the U.S. to Europe is long, it's only going to get longer as they actually are spreading apart from one another.

The evidence for plate tectonics is incredibly strong, and it comes from earthquakes and volcanoes. This is a map of where earthquakes have occurred over a five-year period. You'll notice that it very strongly matches the seafloor topography that I just showed you. The number and strength of earthquakes are highest at plate boundaries. Now, you'll also note that earthquakes can occur away from the plate boundaries.

I take a look, and I live in the eastern United States, and I see there are a significant number of earthquakes that occur in the eastern United States. I'm told that a moderate earthquake in Boston could do more damage than a stronger one in San Francisco because the building codes do not anticipate earthquakes in the Northeast. The same is true for volcanoes. Here is a map of where volcanoes are found on our planet. They are not randomly distributed throughout our planet, but again, they match with the location of the trenches. The subduction of a plate underneath another plate leads to volcanic activity on the continental plate next to it.

But once again, you will notice that the correlation is not perfect. In particular, there are some volcanoes here in the middle of the Pacific Plate. This is Hawaii. Why are there volcanoes in Hawaii? This is as far away from either side of a plate as you can possibly get. Well, let's take a look at the Hawaiian Islands. This is an overhead map of the Hawaiian Islands, and you can see the Big Island actually contains two very large volcanoes. And the northern one is Mauna Kea, and that's dormant, and the southern one is Mauna Loa, and that's still active.

The idea of how you can get these volcanoes out in the middle of the Pacific Plate is something called hot-spot volcanism. There is a magma hot spot underneath the Pacific Plate, and that magma pushes up through the Pacific Plate and forms these volcanoes. As the plate moves across the hot spot, new islands in the chain grow. And you can see here from just the state of Hawaii that the islands continue off to the northwest.

If I take a wider view, you can see that it isn't just the islands that we see today, but there are considerably more islands that have been there previously. And if I look at a topo. map, you can see that if I remove the water, then you can see that there is a huge long chain from Hawaii down here, all the way up across the Pacific Plate. There is at least 70 million years of hot-spot volcanism in the Hawaiian Island chain. There's a new volcano that is forming, just south of the Big Island, and they call it Loihi. That eventually will become a new island in the Hawaiian chain.

So that sort of inspires an intriguing question: What is the history of the plate motion? How have these plates been structured over history? Well, here is how the continents look today, and one thing you will notice is how the bulge of South America on the east coast would fit very neatly into the west coast of Africa. The Mid-Atlantic Ridge has separated these two continents over millions of years.

If you look 60 million years ago, the continents were closer, and you can also note how far India is from the Eurasian Plate. India over the last 60 million years has been a relative speed demon. It's been traveling at up to 15 centimeters per year. If we look back 120 million years ago, this was a time when South America and Africa were in contact. 180 million years ago, the Atlantic Ocean disappears. Matter of fact, you only have one, much larger, ocean, and we have the continent that's called Pangea, one supercontinent composed of all the continents that we know today.

Going further back to 240 million years, the landmasses seem to migrate a little bit south. And going to 300 million years, it still remains mostly one continent, as it continues to migrate south even further. At 360 million years ago, you can see that most of the landmass is below the equator. And when I look at this, I say, "Ah, maybe that north-south dichotomy on Mars, where there's significantly different topography in the south compared to the north, well, I guess it doesn't look so strange any more."

At 420 million years ago, it's no longer one continent. So the supercontinent of Pangea is just a temporary state. North America has broken off. And I also think it's kind of fun to look at this and say, "Hey, look, Alaska, Greenland, and Russia—all of them straddle the equator." So we conclude that the continental plates have been drifting across an interior layer of the planet. I guess "drifting" is one word, but I think "floating" is a better word because that will lead to the idea that's important to discuss next, and that's differentiation.

We all know that oil and water don't mix. If you put oil and water together, and you shake it up, and just let it sit, eventually the water will drop to the bottom, and the oil will rise to the top. The water is more dense, and it sinks to the bottom. For planets, this idea is called differentiation. It's basically the simple gravitational separation of materials of different density. It requires, however, that the body must have been liquid at some point during its formation.

Massive objects will have high temperature and pressure due to gravity. They naturally will become differentiated. What's more important for smaller bodies is the heat generated during formation. The cooling efficiency of a body is really a ratio of its surface area to its volume. And if these things are roughly spherical, the area is proportional to the radius squared, and the volume is proportional to the radius cubed. So the cooling efficiency goes up with size. Smaller objects will cool more efficiently, so you will expect large objects to be differentiated, but the smaller objects will be able to cool away their heat. They won't necessarily become liquid, and we might expect those to be undifferentiated.

The tool that we'll use to explore the interior structure is seismology. And it's simply—the earthquakes trigger seismic waves, and some of those seismic waves will pass along the Earth's surface. These are the ones you feel when you're involved in an earthquake. I did my graduate work at the University of California at Berkeley, and I experienced a number of these. Most of them were kind of fun, a little rolling of the room. But I was also there during the 7.2 Loma Prieta earthquake, and that one scared me a little.

But the other seismic waves, the ones that are useful for exploring the interior, are the ones that are going to go downward and through the interior. The speed of these waves is dependent on the density of the medium. They can also change direction or bounce when there are boundaries between different density mediums. This is like refraction or reflection with light waves. When the wave hits the boundary, it can bounce back, or it can change its direction.

There are two types of waves that we will consider. The first are called primary waves. These are what scientists call acoustic waves, and they propagate like sound waves. There's a compression and rarefaction as it moves along, just like the sound waves that you hear.

The second type are called secondary waves, and these are sheer waves. They're a side to side motion, so this is like when you're wiggling a string or a bullwhip, that side to side, that's a wave passing this way. You'll note that for a sheer wave to work, if I push this piece of a body to the side, it needs to have some connection to the next piece of a body to pull it. Therefore the sheer waves can't propagate through a liquid because if I pull a little bit of liquid to the side, the next bit doesn't go side to side.

So an important property is that the primary waves can pass through solids and liquids, but the sheer waves can only pass through solids. The basic idea is that you know the epicenter of the quake. You measure the strength of the waves on the surface, determine the epicenter of the quake, and then around the globe at the seismic stations, you measure which waves arrive and when they arrive. If, for example, no secondary wave arrives, then you know there must a liquid layer between that seismic station and the epicenter of the quake.

The time of arrival will indicate the average density of the material that it passed through. Now, in this way, you can work to build up a model of the interior. It's not going to be a simple solution, simply because you do have these boundaries, and you have reflections and refractions, but with enough data over many years, we've built up a very good model of the Earth's interior.

Earth's interior structure is considered to be composed of several layers. Basically, the basic layers are crust, mantle, and core. The crust is roughly 60 kilometers thick, only about 1% of Earth's radius. It's composed of the low-density materials, and as you would expect, the low-density materials are the stuff that are going to float to the top of a differentiation process. It really is not very significant because it contains only 0.4% of Earth's mass. It is, however, very significant to us because it's where we live.

One question we have is: Why isn't there more crust? If you look at the materials that we think formed our Earth, we expect there should be 200 kilometers of crust. However, when you do the calculations, at around 60 kilometers the pressure and temperature become so great that the materials of the crust undergo what's called a phase change. They become higher density materials, and the bottom of the crust peels off, they say it "delaminates," and then starts to sink. So the crust is kept at about 60 kilometers thick through this process.

The mantle, the middle layer, is composed of rock and extends about halfway down to the center. It contains two-thirds of Earth's mass. There are three subregions in the mantle. You have the upper mantle, and that contains the asthenosphere. This is a region that acts like a fluid on geologic times. It's still rock, but over millions of years, it actually flows. And there is a slow convection in the asthenosphere, and the lithospheric plates float atop this plastic region.

At the bottom of the upper mantle is what they call the transition region. This is maybe the source of the basaltic magmas. This is where materials that have peeled off from the crust and dropped down become heated enough that they regain their buoyancy and again start their rise back up to the surface.

The lower mantle is solid. It contains the higher-density materials that form under really high pressure. About half the mass of the Earth is in the lower mantle. Below that, we have the core. This is mostly iron, with a little bit of nickel. It contains about one-third of the mass of Earth. It was probably all liquid at one point, but here the Earth has cooled enough that there is an inner core which is solid.

Earth's interior structure—crust, mantle, core—serves as a template for understanding the other rocky planets. If we look at Venus, Venus is expected to have a similar interior structure to Earth. The major difference is that Venus is a one-plate planet. We're not exactly sure why. When we look at Venus, the tectonic features are spread everywhere, not like on Earth, where they were aligned with plate boundaries. There are no plate boundaries on Venus. The entire surface appears to be no older than a few hundred million years. So the tectonics are expected to be compressional sideways and up and down.

Here is a view of the Aphrodite Terra on Venus, and in the center, the red pinkish regions are Ovda Regio and Thetis Regio. These are called tesserae, and these are some of the oldest regions on the surface of Venus. They show this complex deformation and probably were caused by convective motions in the mantle, raising and lowering the crust. This obviously was followed by planet-wide resurfacing because over 80% of the planet is volcanic plains. So for Venus, lava flows are the most significant interior manifestation on the surface.

Mercury has a very high density that indicates that it has a very large iron core. The iron core extends more than two-thirds of the radius of the planet. Originally, we believe that it would have cooled off and become solid, but recent observations have shown that at least it is partially molten. Now, Mercury shows no tectonics, but it does show an interesting feature called fault scarps.

This is an image of a scarp, and the scarp is this line working from in the center down through these craters. This is a geologic fault, and the features like these indicate that a region of the surface has compressed. We believe that perhaps there was even global contraction of the entire planet during its formation. It formed, and it was hot and slowly cooled, and cracked as it cooled.

Mars has a lower density than the other rocky planets. Earth's density is up over 5 grams per cubic centimeter. Mars is more like our Moon, and down around 3 grams per cubic centimeter. It likely has a lower percentage of iron in its composition, and therefore it has a correspondingly smaller core. The core of Mars only extends out to about 30% of its radius, instead of 50% as we see here on Earth.

The expression on its surface in Mars is most noticeable in a region called the Tharsis bulge. In this topographic map, you can see this giant red region over there that has been uplifted several kilometers above the average radius of Mars. In the Tharsis bulge are the giant volcanoes, including Olympus Mons—the largest volcano in the solar system. This likely indicates that in the mantle of Mars, there has been sustained regional motions that have continued to push up. We don't know why this should be, and particularly why it should last for the billions of years that it obviously has in order to produce the extensive volcanism we see here.

I would say that compared to Earth's plate tectonics, the other rocky planets have relatively mild expressions of their interiors in their surface features. If we look at the large moons and the minor planets, we don't see any tectonics. What we see is really a sequence from differentiated bodies down to undifferentiated bodies.

We start with the Moon. The Moon has a relatively low density compared to Earth—about the same as Mars—and it means that it probably does not have a core. This diagram here shows a core, but considering the density, we really doubt that it has a true iron core. The Moon is geologically dead. We see some ancient crustal extension and contraction, stuff that happened during its formation, but it's had no activity for the past 2.5 billion years.

As for the asteroids, we would expect the small bodies of the asteroids to cool efficiently, and the heat of formation would be lost, and therefore they would not be differentiated. Unfortunately, the evidence disagrees with our expectations. We see iron meteorites. They fall to Earth, and these can only come from a differentiated body. They are pure iron on these meteorites. They seem to indicate the complete breakup of differentiated bodies in the asteroid belts, so some asteroids must be differentiated, and we'd expect the largest ones to be.

Here's a diagram of the largest one, Ceres, and it is likely to have a differentiated interior. We have observations that show it has a difference between its crust and an icy layer below that crust. The density of Ceres requires that there be a rocky interior below that. And this diagram doesn't show an iron core, but we believe that to be present as well. Only a few other asteroids are large enough to be differentiated, and most asteroids, we believe, are undifferentiated.

Moving on to the Galilean moons, the most interesting one is Europa. This image here shows what we call the ice rafts of Europa. This is a fractured surface that shows a complicated jumble of ice. In a global view, we see much longer striations, these long lines and fractures that extend across most of its surface. This image looks somewhat like what we see here on Earth at the North Pole. Now, these radial lines are just indicating where the North Pole exists, but these other lines are fractures in the ice above the Arctic Ocean.

We take the same idea and apply it to Europa, and deduce that there might be a subsurface ocean on Europa. So Europa's interior is differentiated, has a metallic core and a rocky interior, but out towards the edge, it is covered with ice and may have perhaps a 10-kilometer-deep ocean. The evidence is consistent with this ocean only existing in the past, so we're not entirely sure that the ocean exists at present.

If we look at the other Galilean moons, in the upper left, we have Io. Then we have Europa in the upper right. We have Ganymede in the lower left. We have Callisto in the lower right. These are diagrams of their interiors as inferred from gravity field and magnetic field measurements. All but Callisto are believed to be differentiated. They have an iron core surrounded by rock, and all but Io have an icy surface. Io, of course, has extensive volcanism and would not have an icy surface.

It is interesting to note that Ganymede and Callisto may also have subsurface oceans. The evidence for this comes from a small variability in their magnetic field. We noticed on Europa that its magnetic field varies just a little bit, perhaps due to the subsurface ocean. Now, to do this, to create a magnetic field, you need an internal conducting fluid because electricity and magnetism are complementary. Wherever you have an electric current, you generate a magnetic field. If a magnetic field changes, you generate an electric current.

So, for Ganymede and Callisto, when we see changes in their magnetic field, the best explanation we can come up with is a subsurface ocean. Now, Europa and Callisto have only tiny magnetic fields. They're really not intrinsic to the body of themselves. They're more a response to Jupiter's magnetic field. Ganymede, however, is drawn with a thick ocean layer because it has a significant internal magnetic field.

Moving on to Saturn, its large moon Titan is also suspected of having a subsurface ocean. It is larger than Mercury, so we expect it to be differentiated and have a rock and ice interior. We also suspect that it has a water and ammonia liquid layer about 50 kilometers below its icy surface.

At the very outer edge of the solar system, Neptune has a large moon, Triton. Triton is thought to be an undifferentiated mix of ice and rock. This is due to its gentler formation in a cold region of the solar system. Also note that ices can stick together under gentle collisions and don't need to heat up that much. Pluto is very similar to Triton in most properties, but it's just a little smaller, so we would expect basically the same interior for Pluto.

Finally, we turn to the giant planets. The interiors for these are mostly determined by modeling, and we find that Jupiter and Saturn are different from Uranus and Neptune. Jupiter and Saturn are composed roughly of solar composition of hydrogen and helium—75% hydrogen, 25% helium. The outer layer of them is gaseous molecular hydrogen. These are the atmospheres that we see, but as you go deeper and deeper into it, that gas becomes really more like a hot liquid. And at about 1 million times Earth's atmospheric pressure, the hydrogen breaks down and becomes a liquid called metallic hydrogen.

In the very center, below this metallic hydrogen, there's a core, and we presume it to be a mass of rock about 15 times the mass of Earth. We have no direct evidence for it, but we need to postulate this mass of rock as a nucleus for the formation of Jupiter and Saturn. Basically, you need enough mass to attract and hold the gases that form the rest of the planet.

Jupiter's density indicates that it is dominated by its liquid hydrogen phase. So though we call it a gas giant, from this viewpoint it's really more a liquid giant. And Saturn's density is actually below that of water, and so by volume, it's mostly gaseous hydrogen.

Uranus and Neptune are significantly smaller. They form farther out in the solar system, and instead of accreting gas, because they didn't have enough mass, they instead accreted "ices." The "ices" in this case mean water, ammonia, and methane. Uranus and Neptune's interiors start with a small rocky core of perhaps a few Earth masses. Then outside that, we have a dominant component that is hot, highly compressed liquid. This is the ices and various chemical species derived from these ices.

The outer 15% of these planets is molecular hydrogen gas. So Uranus and Neptune show a completely different structure from Jupiter and Saturn, and some people refer to these guys as "ice giants," as compared to Jupiter and Saturn as "gas giants," or even "liquid giants."

That completes our journey to the centers of the planets. Now, you may notice that there's one external manifestation that we've not yet discussed. Magnetic fields are generated in the interior of most planets, but the effects are seen on the exterior. In particular, magnetic fields interact strongly with the solar wind. We will discuss magnetic fields in detail in the context of solar storms in a few lectures.

In the meantime, we're just going to work our way out to that region. Our next lecture will move from planetary interiors to planetary atmospheres. We'll see you then. Take care.

Lecture Twelve
Structure and Behavior of Atmospheres

Scope:

The air we breathe is a vital component of most life on Earth. Nowhere else in the solar system can one find the same gases in abundance, and atmospheric pressure elsewhere varies from hundreds of times down to about a billionth of Earth's. The major similarity among atmospheres is the structure of overlying thermal layers where temperatures can both increase and decrease with altitude. The clouds of other planets can be composed not just of water vapor, but also of carbon dioxide, methane, ammonia, and sulfuric acid. Winds can be gentle, with a single global circulation pattern, or they can be many times beyond hurricane speed and show a multitude of turbulent bands wrapping around a planet. In the Venusian atmosphere, one finds a cautionary tale: The small greenhouse effect that warms Earth has become a runaway process that produces hellish conditions on Venus.

Outline

I. Atmospheres across the solar system are diverse in composition and thickness yet similar in structure.
   A. Earth's atmosphere is a very thin layer above its surface, roughly 100 kilometers thick on a planet that is nearly 13,000 kilometers across.
      1. Earth's atmosphere is about 77% nitrogen, 21% oxygen, 1% water vapor, and 1% argon.
      2. While oxygen is the component we emphasize for breathing, nitrogen is much more abundant.
      3. Carbon dioxide plays an important role as a greenhouse gas, yet it is less than 1/30 of 1% of the atmosphere.
   B. The other rocky planet atmospheres are markedly different.
      1. Venus's heavy atmosphere is 96% carbon dioxide and has a pressure about 100 times that of Earth's.
      2. Mars also has a carbon dioxide atmosphere (95%), but it is remarkably thin.

- C. The vertical structure of Earth's atmosphere has four levels.
    1. The lowest level, the troposphere, is dominated by heat absorbed and reradiated by the ground.
    2. Above 15 kilometers, in the stratosphere, ozone molecules absorb ultraviolet light, and the atmosphere gets warmer with altitude.
    3. Thermal inversions also occur at the boundaries of the mesosphere (50 kilometers) and the troposphere (85 kilometers), where temperatures decrease and increase, respectively, with altitude.
    4. The commonly quoted height of the atmosphere, 100 kilometers, does not have a scientific basis and is just an approximate round number that marks the boundary between aeronautics and astronautics.
- D. The atmospheres of Venus and Mars, though of vastly different pressures, show similar vertical structure and extent.
- E. Saturn's moon Titan has an atmosphere that is a bit thicker than Earth's, and which is predominantly nitrogen, with a few percent methane.
- F. Tenuous atmospheres are found on Io, Triton, and Pluto.
- G. The giant planets have deep atmospheres of hydrogen and helium.
    1. Lacking a surface, their heights are not strictly defined, but they span several thousand kilometers and contribute a significant percentage of the mass of the giant planets.
    2. The upper atmospheres are cold, hundreds of degrees below zero, with generally only a single thermal inversion.

II. Clouds, the dominant feature of many atmospheres, are generally condensates of minor constituents.
- A. Jupiter and Saturn's clouds are condensations of ammonia, ammonium hydrosulfide, and water, each at distinct layers in their atmospheres.
- B. Uranus and Neptune have temperatures cold enough to form methane clouds.
- C. Earth's water vapor clouds appear in several varieties at several characteristic heights in the atmosphere.

- **D.** Mars's thin clouds are made of water or carbon dioxide.
- **E.** Venus has thick cloud layers composed of sulfuric acid droplets.

III. Wind patterns on the planets are well organized, though not completely understood.
- **A.** Winds on Earth and Venus show relatively simple thermal circulation.
  1. Venus provides an excellent example of Hadley cell circulation that transports energy from the equator to the poles.
  2. Earth's faster rotation produces three Hadley cells per hemisphere.
  3. The trade winds and jet stream are examples of sustained circulatory patterns on Earth.
  4. Wind speeds on Earth can reach 200 kilometers per hour, but that merely equals the speed of the constant upper jet stream on Venus.
- **B.** The fast rotation of the giant planets leads to a large number of zonal winds.
  1. Lighter regions are called zones, while darker regions are called bands.
  2. The colors of the zones and bands have been observed to change significantly over a period of months.
  3. Wind speeds on the giant planets are strong and stable, ranging up to about 1100 kilometers per hour on Saturn.
  4. Saturn shows a remarkable hexagon feature at its pole.
  5. The winds on Uranus and Neptune are less well studied but show strong winds with only a few zones and bands.

IV. The composition of the atmosphere can play a critical role in determining surface temperature.
- **A.** The greenhouse effect is the trapping of heat within the atmosphere by gases such as water vapor, carbon dioxide, and methane.
- **B.** Earth and Venus show the benefits and drawbacks of warming of the atmosphere.
  1. On Earth, a small warming effect keeps the average temperature at the surface above the freezing point of water.

2. On Venus, a runaway warming pushed the temperature and pressure at the surface beyond the melting point of lead.
3. Earth has about the same amount of carbon dioxide as Venus, but most of it is in rocks.

C. Human behaviors are affecting Earth's atmosphere and climate, with potential for long-term consequences.
1. Since the industrial age, the amount of carbon dioxide in Earth's atmosphere has risen dramatically.
2. The amount of sea ice at the North Pole is visibly shrinking.
3. The scientific evidence for climate change is clear, but how we respond to that evidence is a social and political issue.

Suggested Readings:

Beatty, Petersen, and Chaikin, *The New Solar System*, chaps. 13, 15.

Bennett, Donahue, Schneider, and Voit, *The Cosmic Perspective*, chaps. 10–11.

De Pater and Lissauer, *Planetary Sciences*, chap. 4.

Dessler and Parson, *The Science and Politics of Global Climate Change*.

McFadden, Weissman, and Johnson, *Encyclopedia of the Solar System*, chaps. 7, 9, 15, 20, 25.

Shu, *The Physical Universe*, chap. 17.

Questions to Consider:

1. What are the factors that determine whether or not a planet or moon has an atmosphere?
2. Watching from outside, we can see the banded wind patterns on the giant planets. As there is no surface, how and where would a probe carried along in the atmosphere be able to detect these winds?
3. Could global warming on Mars make its atmosphere hospitable to life?

Lecture Twelve—Transcript
Structure and Behavior of Atmospheres

I would like to start today with an amazing image taken by the Apollo 12 mission. This is an image of a solar eclipse, but it's not by the Moon. This is the Sun being eclipsed by Earth. The Apollo 12 mission flew through Earth's shadow and was able to get this amazing picture. Now, I note that this picture was not staged. They didn't really have to change the mission profile in order to get this picture done because the Apollo missions were generally done during full Moon, so the Moon was at opposition, and flying through Earth's shadow was not really a detour.

Now, it's a beautiful image, but what excites me is looking at that thin crescent of Earth's atmosphere. It's even exaggerated a bit by the backlighting. And it helps me appreciate what a thin blanket Earth's atmosphere really is. It's just a tiny margin that separates us from space, and in this sort of an image, it seems so fragile against all that empty blackness of space. And it's especially so because within that protective layer exists all the life we know of in the universe. On our 13,000-kilometer-sized planet, just a 100-kilometer-thick air bubble is where we exist.

So let's take a closer look at that atmosphere and the other atmospheres across the solar system, and how they are structured, how they behave, and how they affect the prospects of life. Our atmosphere on Earth consists of 77% nitrogen, 21% oxygen, 1% water vapor, and 1% argon. Now, this is not what most folks expect. Oxygen is not the main component of our atmosphere, rather, nitrogen is. Nitrogen is not something we really think of much. Maybe we think about spreading some nitrogen in our garden, but it really is all around us in the air.

Another surprise for a lot of people is that carbon dioxide is not an important constituent. For all that one hears about carbon dioxide being a greenhouse gas, it's really less than 1/30 of 1% of our atmosphere. In contrast, Venus has plenty of carbon dioxide. It makes up 96% of Venus's atmosphere. The other 3% is nitrogen, and then there's some minor amounts of water vapor and argon. Venus has a much thicker atmosphere. The atmospheric pressure at the surface of Venus is about 100 times that of the atmospheric pressure at the surface of Earth.

Earth's atmosphere, in turn, has about 100 times the atmospheric pressure of Mars's atmosphere. Mars has a very thin atmosphere, but it's also mainly carbon dioxide, about 95%, with nitrogen and argon. These bar graphs summarize the compositions of these atmospheres, where you can see strongly that Earth is the odd man out here. Whereas Venus and Mars are dominated by carbon dioxide, Earth is dominated by nitrogen and has enough oxygen for us to be able to breathe.

Another way to look at the atmospheres is through their vertical structure. We say that Earth's atmosphere is about 100 kilometers thick, but really, how do you measure the top of an atmosphere? Earth's atmosphere has four different levels: the troposphere, the stratosphere, the mesosphere, and the thermosphere.

The troposphere extends from the ground up to about 15 kilometers, and it contains 80% the mass of our atmosphere. It is dominated by heat absorbed by the ground and reradiated, so the temperature decreases a few degrees for each kilometer you go up in height in the atmosphere. Well, you've been through the troposphere because if you've taken an airplane flight, airplanes fly at about 10 kilometers, and that's about two-thirds of the way up the troposphere.

Above the troposphere is the stratosphere, and in the stratosphere, ozone molecules absorb the ultraviolet radiation of the Sun, and that causes the atmosphere to heat up. That continues on up to about 50 kilometers, and in there we encounter the mesosphere, and again the temperature then starts to decrease with altitude.

That continues up to about 85 kilometers, where we hit the thermosphere, and the temperature then starts to increase with altitude because the atmosphere is absorbing the high energy radiation from the Sun. The thermosphere includes the ionosphere, the electrically charged region where we see the auroras.

Now, the height of the atmosphere, we say, is about 100 kilometers. Now, that's almost the top of the mesosphere, but not quite. Actually, here's a secret for you. It isn't really derived from the properties of the atmosphere. It basically was a calculation as to what is the highest altitude a plane could fly, and basically, it's the boundary between what you would call aeronautics and astronautics. They got a number around 100 kilometers and they rounded it off to 100 kilometers because it's a good, honest number. So you are not officially an astronaut unless you've crossed the 100-kilometer barrier.

Here is a graph of the atmospheric temperature with height, and the blue line here is that of Earth, and you can see how it decreases through the troposphere, increases through the stratosphere, decreases again through the mesosphere, and increases in the thermosphere. We can compare these regions to those on Venus and Mars. For Venus, we define the same levels, but you can see that the troposphere, the region where it's decreasing with temperature, goes up much, much higher. It goes up to about 65 kilometers.

Now, I've heard the region on Venus between 65 kilometers and 95 kilometers called both the stratosphere and the mesosphere, so I'm not really sure which to call it. We'll just call it the middle atmosphere. And then the thermosphere begins around 95 kilometers. And so while the atmosphere of Venus is 100 times heavier than that of Earth, its height is about the same—95 kilometers versus about 85 for Earth.

And surprisingly, the same is true for Mars. Its upper atmosphere starts at about 110 kilometers. Its middle atmosphere extends down to about 45 kilometers, and the lower atmosphere goes from there to the ground. For three atmospheric pressures of vastly different scales, it is surprising that they have roughly similar vertical extent.

Also similar and surprising is the atmosphere of Saturn's large moon Titan. Here's an image of Titan, with it being backlit by the Sun, somewhat like that solar eclipse image I showed you. And it's surprising, first of all, that Titan even has an atmosphere. The other moons are the same size: Ganymede, Callisto, Europa, our Moon, as well as Mercury, which is smaller, by the way, than Titan. They do not have atmospheres.

Secondly, it's surprising that its atmosphere extends to about 600 kilometers away from the moon. This is six times the extent of the atmospheres of the rocky planets. And third, it's surprising that for a moon, its atmospheric pressure is 1.5 times that of Earth. Now, considering the wide range of pressures on Venus and Mars, this sort of counts as a similarity as well because Venus is 100 times larger and Mars is 100 times smaller, so being only 1.5 is really quite similar to Earth.

Another similarity is that it is predominantly a nitrogen atmosphere. It is 98% nitrogen and about 2% methane. Methane reacts with the Sun's ultraviolet radiation and forms a very dense smog layer high up in Titan's atmosphere. This is why you can't see through it. What you are seeing in this yellow region over here on this image is the smog layer high up in Titan's atmosphere. The Huygens probe that landed on the surface of Titan could not see the Sun—so finally, out in the solar system, we have found worse smog than exists in Mexico City or Los Angeles.

A couple of other moons have rather tenuous atmospheres. Io has an atmosphere that derives from its volcanic emissions. It's basically sulfur dioxide that has been spewed out by the volcanoes, or actually sublimated from the surface after the volcanic spewings condense on the surface. It is one-billionth the pressure of Earth, and even that is lost to space relatively quickly. If you could suddenly turn off all the volcanoes on Io, eventually its atmosphere would disappear.

Neptune's large moon Triton has an atmosphere of nitrogen and methane, just like Titan. Triton is like Titan. Sometimes you get those words mixed up. However, Titan's atmosphere is over 10,000 times more dense than Triton's. Triton is a very, very thin wispy atmosphere that rises about 10 kilometers above its surface.

Pluto's atmosphere is essentially the same as Triton, and due to its elliptical orbit, Pluto's atmosphere isn't permanent. It will actually freeze out as it gets further from the Sun. Now, the New Horizons mission is headed toward Pluto, and it will arrive there in 2015. I hope there's enough atmosphere left that they can study it when they get there.

The giant planets have deep atmospheres of hydrogen and helium, and they descend down to very high temperatures and pressures. And so instead of asking, "Where's the top here?" you can also ask, "Where's the bottom?" Well, since there is no surface on these giant planets, the thickness is irrelevant. It goes down for thousands of kilometers. What really is relevant is that the atmospheres are a significant percentage of mass of the planet. On the rocky planets, the atmospheric mass is a tiny, tiny percentage of the total mass of the planet, but here in the gas giant planets, it's a significant percentage.

Now, as with the discussion of their interiors, Jupiter and Saturn fall into a little different category than that of Uranus and Neptune. Here are the temperature curves for Jupiter and Saturn. At the zero point here, I will note it is arbitrarily set at the pressure of 0.1 times Earth's atmosphere, and the y-axis shows you the altitude in kilometers above or below that arbitrary zero point.

The first thing you should notice is that these are very cold atmospheres. The temperatures on the x-axis are in the Kelvin scale, where zero is the absolute zero. The degree range is the same as Celsius, but with water freezing at 273°K. So if you take these temperatures and subtract 273°, you get temperatures on the Celsius scale.

The warmest temperatures on these plots are basically normal temperatures on Earth. And when you look at the tops of the atmospheres, they're about −200°. You see the same pattern as we saw for the rocky planets. The temperature decreases with height as you go through the troposphere, and then it increases with height as you get through the thermosphere.

The image here also shows the major cloud layers on Jupiter and Saturn. The different substances in this atmosphere will condense out at different temperatures and pressures. There are three distinct cloud levels in the atmospheres: one of ammonia ice, one of ammonium hydrosulfide, and one of water ice. So, when you're looking at different clouds on Jupiter and Saturn, you're really looking to different depths and temperatures within the atmospheres.

For example, here's a picture of Jupiter from the Hubble Space Telescope. Note that it has many brown and white bands. Now, the top is brown, and then you've got a couple of brown bands up in the northern hemisphere, and some brown bands and a little bit of brown towards the south pole. Contrast that to an infrared telescope image of it that sees the heat. When looking at the heat signature, those brown bands then become very bright in the infrared. What is happening is you're seeing deeper into the atmosphere and seeing higher temperatures through those brown bands. The brown bands are the lower clouds at higher temperature within Jupiter's atmosphere.

These are the atmospheric temperature curves for Uranus and Neptune. The situation on Uranus and Neptune involves the same ideas we've already discussed, except we're dealing with even colder temperatures. You can see that to the depths we've probed in these two planets, the temperatures are still 100° to 150° below freezing—and at their coldest, near the tops of the atmosphere, they're over 200° degrees below zero.

This allows clouds that we would never see on any other planet, methane ice crystals, because methane doesn't freeze until −180°C. Here is an image that shows you these clouds. These are the white streaky features here on the planet Neptune. Note that the blue color of Neptune is also due to methane. Methane absorbs red light, and only blue light is reflected, and that gives Neptune its beautiful blue color.

This reminds you a little bit of what we see on Earth. On Earth, we see blue skies with white clouds. This is visually similar, but physically, it's entirely different. Earth clouds, as you probably know, are made of water vapor. They occur at many different levels, but they actually have distinct levels. And again, if you've taken an airplane ride, you've probably noticed that sometimes you have to pass through the cloud deck, and it can be one type of weather beneath the cloud deck, and a totally different type of weather up above the cloud deck, but there's a distinct level of clouds that you pass through.

Also, Earth's sky is blue because the air molecules are scattering blue light better than they scatter red light. Now, when you ask people, "Why is the sky blue?" sometimes they'll give you a crazy answer like, "Oh, because it reflects the water." And that's just wrong. Okay? I've just got to get rid of that misconception. If that were true, why would the sky be blue in Kansas when there's no water around? The sky is blue because the air molecules preferentially scatter blue light better than they scatter red light.

Now, Mars also has some wispy clouds. Here is an image of the clouds on Mars, and you can see them in the northern hemisphere, up towards the north pole, these very wispy white things. They have lower clouds that are made of water vapor and the higher clouds that are cold enough to be carbon dioxide ice.

Venus's clouds, well, I think Venus's clouds are probably the coolest clouds anywhere in the solar system because get this—they are actually made of sulfuric acid droplets. That's right, sulfuric acid clouds. How cool is that? They have a very thick cloud layer, about 50 kilometers up in the atmosphere. It's continual clouds for about 10 kilometers, a 10-kilometer-thick cloud deck. So if you're on the surface of Venus, just like on the surface of Titan, you can't really see the Sun.

This also means that Venus is rather featureless when seen in visible light. The image I'm showing here uses ultraviolet light to sort of pierce through some of the clouds and be able to see the patterns in it. That sort of chevron pattern, you see the sort of Y-shaped pattern? That leads us to our next topic, which is that of winds.

The basic idea for Earth and Venus is something we call Hadley cell convection. It's really simple. You already know it. It's basically hot air rises, cold air descends. Straightforward enough, right? Well, apply that to Earth. Where is it hottest? Along the equator. So hot air will rise at the equator. Where is it coldest? At the poles. Cold air will descend at the poles. So you have a simple natural circulation pattern from rising at the equator, streaming up to the poles, and descending at the poles.

On Venus, it has a relatively slow rotation. Venus rotates in about 243 days, and that allows it to have one Hadley cell per hemisphere. This is a beautiful example of Hadley cell circulation. And when you've got these circulation patterns going up, the winds, the jet stream, on Venus drive that chevron pattern. So, take a Hadley cell circulation and then start to move it with jet stream winds, and you get that chevron pattern. I note that these winds are about 200 miles per hour. While Venus rotates in 243 days, the winds circulate around the planet in only 4 days.

On Earth, we have faster rotation. Instead of producing one Hadley cell, we actually produced three cells. Near the equator, we're familiar with the trade winds, and they blow from east to west along the equatorial latitudes. At the mid-latitudes, such as that of most of the United States and Europe, the winds blow from west to east. And then a third cell near the poles, again blow back in the opposite direction, east to west. I note that Antarctica has become a really great place for balloon-borne science because you can launch the balloon, it will follow its winds all the way around the pole and come back towards the launch area, where you can recover it.

In the U.S., we see these different trade-wind motions in hurricanes. The familiar pattern is that a hurricane will form out in the southern Atlantic Ocean. It will move westward towards the U.S. It will turn up the coast and move back eastward as it dissipates. It's born in the trade winds, and then it dissipates in the westerlies.

The wind speeds on Earth or in these hurricanes: A level 1 hurricane is 75 miles per hour. They go all the way up to a level 5 hurricane. That's 150 miles per hour. On Earth, we have recorded wind speeds in excess of 200 miles per hour. Is that amazing? Well, no, not when you view it in the context of the solar system, because the fastest speeds on Earth are equivalent to the 200 miles per hour jet stream that is seen on Venus all the time. And that's to say nothing of what we see on the giant planets.

Jupiter is 10 times larger than Earth, but it rotates in half the time. Jupiter spins on its axis once every 10 hours. This produces a large number of zonal winds. These are the banded structure that we see in the planet. The lighter regions are called zones, and the darker regions are called bands. Like on Earth, one region is blowing west, and the region adjacent to it is blowing east, and the next one is blowing west, and the next one is blowing east. You can see from the image there are about a dozen or so of these regions. And sometimes we think of these bands as permanent features, but they can change.

Here are two images from the Hubble Space Telescope, and what I want to point out to you is on the left-hand image in March, you can see this white zone in the north. When we looked at it again in June, that white zone had become a dark band. Likewise, in the southern part of Jupiter, there's a sort of turbulent region that is dark in June, but looking back into March, it had an inverted structure and was a lot more calm in March. So, while a tiger can't change its stripes, it appears Jupiter can.

The wind speeds on Jupiter are relatively calm. Most of them are about 40 miles per hour, especially as you go towards the poles. The central band reaches about 200 miles per hour. Here is a plot of the Jupiter wind speeds with latitude. The x-axis is plotted in meters per second, but it's relatively simple. You can just multiply it by 2.2, and you'll get miles per hour.

The fastest winds in the solar system are on Saturn, and here in comparison to Jupiter, it's the same plot for Saturn. And you can see where Jupiter's speeds reach about 200 miles per hour. Saturn's speeds reach 1100 miles per hour. That's five times the maximum ever recorded on Earth. And this isn't just blowing once or twice. This is blowing all the time, every day.

Now, when you think about the giant planets and wind speeds, it's sure to bring up a question: wind speed relative to what? On Earth, it's relative to the ground. On Saturn, it's relative to the internal rotation rate. However, recent Cassini observations have made us unsure exactly what the internal rotation rate of Saturn is, so although this is a record in the record books, we may have to put an asterisk next to it because it could be revised later. Still, no matter what we discover about the internal rotation rate, these winds are really fast.

Now, when we look at Saturn, we can see that these winds produce the same banded structure that we see on Jupiter, although it's much more muted than we see on Jupiter. One unusual feature of Saturn is found at its poles. Remember how we had circular polar winds on Earth? The same thing appears on Saturn, but here it's in the shape of a hexagon. Don't ask me why. We just don't know, but you can see from this animation that it's a stable pattern. It was seen 30 years ago by Voyager and was seen again recently by Cassini.

The situation on Uranus and Neptune is a little less clear because we have fewer observations. Uranus and Neptune have only been visited by one satellite, Voyager 2. Here are the winds on Uranus and Neptune, and you can see what is one thing that's interesting about them in comparison to Jupiter and Saturn is their equatorial jet streams flow west instead of flowing east. These jet streams are slower than the internal rotation period. The pole flow, as you can see, is more towards the east.

Now, in preparing for this lecture, I looked up things on the Internet. I found a lot of references that say, "Neptune has the fastest wind speeds in the solar system," and they were claiming 1500 miles per hour winds on Neptune. And I couldn't reconcile it. And the only way I could reconcile it is if you look at Neptune, and you take the peak of the eastward motion and the peak of the westward motion, if you stretch that, you could probably get on the order of 1500 miles per hour winds. It seems that there was a press release in 1989 that quoted this as the measurement, but it ignores the internal rotation rate of Neptune, so it really isn't there.

Neptune's strong winds produce that same banded structure, and it's somewhat evident, like it's even subtler than it is on Saturn. On Uranus, if we just look with visible light, you don't see much, but if you enhance the image—and this right-hand image is the exact same as the left-hand image, but only enhanced—you can find the banded structure. It also shows that really well when looking in infrared.

The last topic that I would like to discuss today is the balance of atmospheres, specifically the heating and cooling. Here's an image that discusses the heat balance of an atmosphere. The atmosphere absorbs energy from the Sun, sunlight is reflected back by clouds. Some is absorbed by the atmosphere, like the ozone layer absorbs ultraviolet light; some is absorbed by the surface. And then the surface reradiates that energy in infrared light. Some of that reradiated energy is also absorbed by the atmosphere, and some is radiated back to space.

The major point of this balancing argument is that the inflow of energy from the Sun must be balanced by the outflow of energy from reflection and reradiation of that energy. If not, the planet will either be heating or cooling. Now, the greenhouse effect is the trapping of infrared radiation within the atmosphere, and the major molecules responsible for this are water vapor, carbon dioxide, and methane. The extra heat absorbed by these molecules acts like a blanket and keeps Earth's surface warmer.

Here is that same temperature plot that I showed you before, but now I have an arrow here, showing you where Earth's temperature on the surface would be if there was no greenhouse warming. Earth's average surface temperature, without greenhouse, would be −16°C or, 3°F. The oceans would mostly be frozen, so a little bit of greenhouse warming is a good thing.

Now, Venus is just a little closer to the Sun, but because the amount of energy is proportional to the orbit radius squared, it actually absorbs twice as much energy from the Sun as Earth. It has a 96% carbon dioxide atmosphere, which traps a lot of infrared radiation, so that caused the surface to heat up, and heat up, and heat up until enough infrared radiation could escape to balance the incoming light from the Sun. This turned into a runaway greenhouse effect.

And if you look, the surface temperature of Venus is way out here at 475°C. That's 900°F. If you add in the fact that its atmospheric pressure is 100 times that of Earth, it is hot and dense enough to melt lead. The Venera probes that actually landed on the surface of Venus only lasted for about 20 minutes. Eventually they probably just melted. And if you look at the plot without a greenhouse effect, Venus would be colder than Earth.

Now, we also have Mars plotted here, but its atmosphere is too thin. Even though it has plenty of carbon dioxide, it's too thin for much of a greenhouse, and its temperature without a greenhouse is pretty much the same as its temperature is with a greenhouse effect. So Venus is a caution. You can have too much of a good thing. So how does this relate to global warming?

The point is that the Earth has about the same amount of carbon dioxide as Venus, but on Earth, most of it is locked up, mostly in the limestone deposits on the seafloor. If we look in the ice cores, we can see how much carbon dioxide has been in the atmosphere for the last 400,000 years. Here is a plot from the Vostok ice core from Antarctica, and it shows the carbon dioxide record for the last 400,000 years has ranged basically between about 190 and 290 parts per million. Again, carbon dioxide's a very small part of the atmosphere, and here's its historical range.

The low points correlate when Earth was experiencing ice ages, and the high points with interglacial periods. We happen to be in an interglacial period right now. If we look at a more recent historical record, we can see that over the last 1000 years there's a sharp spike in the carbon dioxide from the Industrial Revolution. And if we look at current data, over the last 50 years, we show a continued consistent rise. The level is now 380 parts per million. This is the same jump since 1750 as from 1750 back to the last ice age.

The big question that comes from this is: Will the increased CO_2 levels change the temperature of Earth's atmosphere? Evidence says that this is indeed happening. Here's a movie that shows you the sea ice at the North Pole over the last 20 or 30 years. It begins in 1979 and slowly steps through every year. You can see it growing and shrinking, and growing and shrinking. These measurements are made in September of every year, which is actually the seasonally smallest time for the sea ice.

You can see in 2007 that it's quite small. Just to remind you, here's the sea ice in 1979, and here is the sea ice in 2007. It's the smallest ever recorded in September 2007. For the first time in history, the Northwest Passage above Canada opened up for a brief time. The trend has been downward, and if it continues, in a few decades the polar cap will melt completely during the summer.

Now, the scientific evidence of global warming is real, but the atmosphere is a very complex system. There are a lot of feedback processes, but it's certainly clear that something is changing, and Venus shows us what a worst-case scenario looks like. On the other hand, how we react to this scientific evidence and what we plan to do about it is a social and a political issue, but I think it's something that we should seriously consider.

I'll leave it there for today. And next we'll talk about the weather, but not just any weather—some really serious, extreme weather. We'll see you then.

Timeline

B.C.E.

- c. 350 .... Aristotle creates philosophical view of perfect and immutable heavens.
- c. 300 .... Euclid's *Elements* encapsulates the knowledge of geometry.
- c. 270 .... Aristarchus estimates the relative distances to and sizes of the Moon and the Sun.
- c. 250 .... Eratosthenes estimates the size of Earth.
- c. 150 .... Hipparchus creates a detailed Earth-centered solar system model that is a precursor to the work of Ptolemy.

C.E.

- c. 150 .... Ptolemy publishes a compendium of astronomical knowledge, including what would become the standard Earth-centered solar system model for the next 1500 years. Arabic astronomers later call this work the *Almagest*.
- c. 850 .... Ptolemy's *Almagest* is translated from Greek into Arabic.
- 1175 .... Ptolemy's *Almagest* is translated from Arabic into Latin and reemerges in European studies.
- 1515 .... First printing of Ptolemy's *Almagest*.
- 1543 .... Nicolaus Copernicus publishes his heliocentric model of the solar system in *On the Revolutions of Heavenly Spheres*.

1573	Tycho Brahe's observations of the supernova of 1572 show that the Aristotelian idea that the heavens are unchanging is incorrect. His observations of a comet in 1577 support the same conclusion.
1588	Brahe publishes his partially Earth-centered, partially Sun-centered model of the solar system, known as the Tychonian model.
1601	Johannes Kepler succeeds Brahe in his post and gains access to his observations.
1609	Kepler publishes *Astronomia nova*, which contains two of his three empirical laws of planet orbits.
1610	Galileo publishes the first telescopic observations in *Starry Messenger*. Moons of Jupiter show that not everything revolves around Earth.
1616	Copernicus's *On the Revolutions of Heavenly Spheres* is suspended until corrections can be made. Corrections allowing its reading are issued in 1620.
1619	Kepler publishes *Harmonices Mundi*, which contains his third empirical law of planet orbits.
1627	Kepler publishes the *Rudolphine Tables*, which predict planet positions based on the heliocentric model of the solar system.
1633	Galileo is tried by the Inquisition for promoting heliocentrism as truth in *Dialogue Concerning the Two Chief World Systems*.

1687	Isaac Newton publishes *Philosophiae naturalis principia mathematica*, known simply as *Principia*, which includes his theory of gravity.
1758	As predicted by Edmund Halley, a comet appears with the same orbit as bright comets of 1531, 1607, and 1682.
1772	Titius-Bode rule on the spacing of planet orbits is published.
1781	William Herschel discovers Uranus.
1801	Ceres is discovered and called a planet. Discoveries of Pallas, Juno, and Vesta, also called planets, follow in 1802 to 1807.
1835	Comet Halley returns, reaching perihelion one day later than expected and fueling suspicions about a planet beyond Uranus.
1846	Urbain Le Verrier and John Couch Adams provide mathematical predictions of a planet beyond Uranus. Neptune is discovered by Johann Galle based on Le Verrier's prediction.
1847	The planets between Mars and Jupiter are, by general consensus, demoted to the status of minor planets, or asteroids.
1859	Le Verrier predicts a planet or swarm of planets interior to Mercury in order to explain the precession of its orbit. One such planet, called Vulcan, is observed once but never again.

1895	Percival Lowell publishes the first of three books on Mars that envision a civilization of Martians. His works are based on mistaken observations of canal structures on Mars.
1916	Albert Einstein publishes his paper on the theory of general relativity. The precession of the perihelion of Mercury is explained.
1930	Clyde Tombaugh discovers Pluto.
1951	Gerard Kuiper notes that a massive Pluto would scatter objects outward. Though this idea is incorrect (because Pluto is not massive), his name is later attached to this region, called the Kuiper Belt.
1957	The first artificial satellite, Sputnik 1, is launched. The space age begins.
1958	Explorer 1 detects the radiation belts surrounding Earth. NASA is founded.
1962	The first interplanetary probe, Mariner 2, is launched on a mission to Venus.
1966	First robotic Moon landing.
1969	First human Moon landing.
1972	Viking 1 and 2 land on Mars.
1975	Venera 9 sends back first photographs from the surface of Venus.
1977	Voyager 1 and 2 are launched.
1978	James Christy discovers Pluto's first moon, Charon.
1979	Voyager 1 and 2 fly past Jupiter.

1980	Voyager 1 flies past Saturn. Voyager 2 does so the following year.
1985	Mass of Pluto is measured to be tiny, 0.2% that of Earth.
1986	Voyager 2 flies past Uranus; first flyby of a comet to see the nucleus (Comet Halley).
1989	Voyager 2 flies past Neptune.
1990	Magellan spacecraft arrives at Venus.
1992	First Kuiper Belt object (not counting Pluto and Charon), 1992 QB1, is discovered; Pope John Paul II endorses a church commission's findings that Galileo should not have been condemned.
1995	Galileo satellite arrives at Jupiter.
1997	First automated surface rover, Mars Pathfinder, lands on Mars; Mars Global Surveyor reaches orbit around Mars.
2003	A Kuiper Belt object larger than Pluto, Eris, is found.
2004	Cassini satellite arrives at Saturn.
2005	Huygens probe lands on Saturn's moon Titan.
2006	International Astronomical Union defines "planet"; Pluto demoted.
2006–2020	Launch of New Horizons mission to Pluto (to arrive 2015) and the Kuiper Belt (to arrive 2015–2020).

Glossary

accretion: The buildup of the Sun or planets by the gradual addition of material.

acid rain: Rain with a small, but unusually high, percentage of dissolved sulfur dioxide.

amplitude: A measure of the height of a waveform. For a wave in the ocean, the amplitude is the distance the wave rises above the ocean's surface.

analemma: The figure-eight shape created by recording the position of the Sun at the same time every day over a year.

Andromeda Galaxy: The large galaxy nearest to our Milky Way Galaxy.

angular momentum: A conserved quantity of spinning or orbiting motion around a pivot point. The product of mass times velocity times the distance to the pivot point.

aphelion: The point in an orbit farthest from the Sun.

asteroid: A small to tiny solar system object made mostly of rock.

asteroid belt: The region between the orbits of Mars and Jupiter where most asteroids orbit.

astrometry: The precise measure of positions on the sky.

astronomical unit (AU): The average distance of Earth from the Sun; about 150 million km or 93 million mi.

Bok globule: A small, dense, and dark region of an interstellar cloud where stars may form.

bow shock: A shock is a sudden jump in pressure in a gaseous or liquid medium. A bow shock occurs on the leading edge when an object is traveling through the medium at high speed (such as a supersonic plane in Earth's atmosphere) or the medium is flowing around an object at high speed (such as the solar wind passing over Earth's magnetic field).

brown dwarf: A body larger than a planet but smaller than a star. It is large enough for deuterium fusion to occur in its core but not large enough for hydrogen fusion. Brown dwarfs are roughly 15 to 70 times the mass of Jupiter.

Cassini Division: A prominent gap between the A and B rings of Saturn.

CCD: An acronym for "charge-coupled device." CCDs are the primary digital devices used for taking images, whether in cell phones or at astronomical observatories.

celestial equator: A projection of Earth's equator outward onto the sky.

Celsius: A temperature scale in which 0° is the freezing point of water (equal to 32°F) and 100° is the boiling point of water (equal to 212°F).

Centaurs: Small icy bodies, like Kuiper Belt objects, that orbit among the giant planets inside of Neptune.

chasma: A deep, narrow canyon created by geologic faulting.

Chicxulub impact: A very large impact crater on the Yucatán Peninsula that dates to 65 million years ago and provides strong evidence that an impact event ended the era of the dinosaurs.

chondrule: A millimeter-sized spherical inclusion found in the oldest meteorites (called chondrites). Chondrules were formed by the heating of minerals to high temperatures followed by rapid cooling.

chromosphere: The region of the Sun's atmosphere just above its surface. About 1000 km thick.

coma: The large gaseous cloud around the nucleus of a comet when it passes near the Sun. Also called the "head" of the comet.

comet: A small to tiny solar system object made mostly of ices, with some rock. When near the Sun, comets develop a gas and dust cloud around them (the coma) that streams away as long tails. See also **long-period comet** and **short-period comet**.

constellation: One of 88 regions covering the entire sky. The bright stars within each region have been variously interpreted as representing people, animals, and objects.

convective zone: The region of the Sun's interior where energy is transported by the overturning motions of gas. See also **radiative zone**.

core of the Sun: The innermost region of the Sun, where energy is generated by nuclear fusion.

corona of the Sun: The outer region of the Sun's atmosphere, extending for millions of miles into interplanetary space.

corona on Venus: An oval-shaped volcanic uplift.

coronal loop: A loop of the Sun's magnetic field arching from its surface through the corona and back to the surface.

coronal mass ejection: The explosive release of billions of tons of hot plasma from the Sun into interplanetary space.

coronography: An observing technique in which the light of a star is blocked in order to reveal fainter details around it.

crater: A circular hole in a planetary surface caused by the impact of an object from interplanetary space.

crust: The thin outer region of Earth's interior.

debris disk: The protoplanetary disk around a newborn star after the low density material has been swept away by the intense radiation and strong stellar winds in the final stages of star formation.

declination: The angular measure of distance on the sky from the celestial equator. See also **right ascension**.

density: The amount of mass per unit volume, generally measured as weight per cubic centimeter (cc). Water is 1 gram per cc, while rock is several grams per cc.

deuterium: Hydrogen with a neutron added to its nucleus. Also called "heavy hydrogen."

differentiation: The gravitational process by which heavier materials sink to the center of a molten or liquid body.

disconnection event: The detachment and subsequent reforming of the ion tail of a comet.

Doppler shift: A change in the observed wavelength (of light or sound) due to relative motion of the emitter and observer.

dust devil: A small swirling tornado-like storm on Mars.

dust tail: The curved tail of a comet carried back by the pressure of the solar wind.

dwarf planet: In 2006, the IAU defined a "dwarf planet" to be an object that satisfies the criteria to be a planet except that it does not dominate its orbit. This definition is relatively useless in practice, as the objects which fit these criteria are simply the largest members of the more general categories of asteroids and Kuiper Belt objects. Such a cross-category grouping is not a true family of objects.

dynamo: A method of producing a magnetic field in the interior of stars and planets that is based on rotation and convection of a conducting fluid.

eccentric circle: A geometrical construct for solar system orbit models in which a planet executes circular motion but the observer is located away from the center of the circle. The planet's motion is uniform, but the off-center observer sees it speed up and slow down during its orbit. See also **equant**.

eccentricity: The measure of the elongation of an elliptical orbit. Zero is perfectly circular, and one is infinitely stretched out to a line. The eccentricity of Earth's orbit is 0.017, while that of Comet Halley's orbit is 0.967.

ecliptic: The path of the Sun across the sky. Equivalent to the plane of Earth's orbit around the Sun.

ejecta: The material excavated during an impact that is thrown from the crater.

electromagnetic spectrum: The entire range of light waves including gamma ray, X-ray, ultraviolet, visible, infrared, microwave, and radio light. Electromagnetic refers to the fact that light is created by oscillating electric and magnetic fields.

elongation: Angular distance on the sky from the Sun. Mercury and Venus are most easily seen at greatest elongation.

epicycle and deferent: A geometrical construct for solar system orbit models in which the orbit circle, the deferent, is combined with a suborbit circle, the epicycle. The planet orbits along the epicycle, and the center of the epicycle orbits along the deferent.

equant: A geometrical construct for solar system orbit models in which a planet follows a circular path, but the motion along that path is uniform with respect to a point away from the center of that path.

When seen from the center of the path, the planet speeds up and slows down during its orbit. See also **eccentric circle**.

equinox: One of the two points during Earth's year, in the spring and fall, when day and night are of equal length.

extrasolar planet (also called **exoplanet**): A planet orbiting a star other than the Sun.

extremophiles: Organisms found on Earth that live in extreme conditions, such as extremes of heat, salinity, or darkness.

filament: A long streamer of gas in the corona of Sun, often many hundreds of thousands of kilometers long. A filament is a streamer seen projected onto the face of the Sun, whereas a **prominence** is a streamer that is seen extending off the edge of the Sun.

fusion: A nuclear process that combines atomic nuclei and particles to create heavier nuclei while releasing energy.

galaxy: A vast collection of many millions to billions of stars, along with huge clouds of gas and dust.

Galilean transformation: A change of reference frames for measurements in which the two reference frames are in uniform linear motion with respect to one another.

gamma ray light: The shortest wavelength and highest energy light of the electromagnetic spectrum.

general relativity: Einstein's reinterpretation of gravity not as a force but as mass warping space. It is required to explain some gravitational effects observed around very massive objects.

geocentric: Centered on Earth.

giant planet: One of the four large planets of the solar system: Jupiter, Saturn, Uranus, and Neptune.

granules: Millions of small (1000-km) cell-like divisions of gas in the photosphere of the Sun. They represent the tops of convective cells from within the convective zone.

gravitational lensing: The warping and amplification of light as it passes through the gravitational field around a massive object. It is an effect due to general relativity.

greenhouse effect: A nickname given to the trapping of heat within an atmosphere due to the absorption of infrared light by molecules such as water vapor, carbon dioxide, and methane.

habitable zone: The region around a star where temperatures are appropriate for liquid water to exist on the surface of a planet.

heavy bombardment era: The time, lasting about a billion years, in the early solar system when impacts were a dominant process shaping the surfaces of planets and moons.

heliocentric: Centered on the Sun.

heliopause: The outer boundary of the solar wind, where its pressure equals that of the gas in interstellar space.

helioseismology: The study of waves pulsing through the Sun in order to learn about the solar interior.

helium-3: Helium that is missing a neutron in its nucleus.

Herbig-Haro object: A forming star, usually hidden inside a dark disk, with twin jets of oppositely directed emission.

hot Jupiter: An extrasolar planet with mass comparable to Jupiter but found in an orbit very close to its star.

hot Neptune: An extrasolar planet with mass comparable to Neptune but found in an orbit very close to its star.

hydrology: The water cycle of evaporation, condensation, and precipitation on Earth; the discipline studying those processes.

hypothesis: An explanation that sounds plausible and fits some observations but has not been fully tested.

IAU: International Astronomical Union. The IAU is the official organization for the naming of solar system objects.

ICRS: International Celestial Reference System. The standard for notating positions on the sky that was adopted in 2000. It is based on the positions of about 200 distant radio sources and stays fixed in space over time. Previous coordinate systems, based on the position of the Sun at the spring equinox, had a slow shift in coordinates over time.

igneous rock: Rock that cooled from a molten state.

infrared light: Light waves with longer wavelengths and less energy than visible light.

interferometry: The technique of combining observations from multiple telescopes in real time to achieve the resolution of a larger telescope.

ion tail: The straight tail of a comet consisting of charged particles carried away by the magnetic field of the solar wind.

ionization: The process of heating an atom or molecule until it is hot enough to lose electrons (i.e., become ionized).

irregular moon: A moon of a planet that is believed to have been captured by the planet's gravity after formation.

Kelvin scale: An absolute temperature scale whose zero point is equivalent to $-273.15°C$ or $-459.67°F$.

K-T extinction: The mass extinction at the end of the age of the dinosaurs, on the Cretaceous-Tertiary boundary, about 65 million years ago.

Kuiper Belt: A region of the solar system beyond Neptune (approximately 30 to 50 AU) containing many thousands of small, icy worlds, including Pluto.

Large Magellanic Cloud: The larger of two prominent dwarf galaxy companions of the Milky Way Galaxy. The LMC, as it is called, is about 170,000 light-years away and is only visible to observers in the southern hemisphere on Earth.

long-period comet: A comet that orbits the Sun in more than 200 years.

luminosity: The energy output of a star, and a measure of its brightness.

magma: Molten rock below the surface of Earth.

magnetosphere: The region of influence of a planetary magnetic field.

main sequence star: A star in the main phase of its life, during which it is powered by nuclear fusion of hydrogen into helium.

mantle: The middle region of Earth's interior.

mare: Smooth volcanic plains on the Moon. The term comes from the Latin word meaning "sea."

meridian: The projection of the local north-south line onto the sky.

mesosphere: The third layer in Earth's atmosphere, in which the temperature decreases with height.

metallicity: A measure of the amount of "metals" (elements heavier than hydrogen and helium) in a star.

metamorphic rock: Rocks formed from igneous or sedimentary rocks subjected to high pressure and temperature.

meteor: A bright streak in the sky created by an object from interplanetary space glowing as it passes through the atmosphere.

meteor shower: A period of high meteor activity associated with Earth's passage through the orbit of a comet.

meteorite: The remains, or a piece, of an object from interplanetary space when found on the surface of the planet.

microlensing: The gravitational lensing observed when a star passes directly in front of another star from our point of view.

microwave light: Light that has shorter wavelengths than radio waves and longer wavelengths than infrared light.

Milky Way: Our galaxy, which appears to us as a band of light across the sky.

momentum: A conserved quantity of motion. It is the product of mass times velocity.

nebula: A large cloud of gas and dust within a galaxy.

neutrino: A subatomic particle that has no charge, has almost no mass, and interacts with matter extremely rarely.

neutron star: The extremely dense remains of a star that has exploded as a supernova. It is like an atomic nucleus as massive as the Sun, and it is composed entirely of neutrons.

nucleus of a comet: The solid, icy, and rocky body of a comet, from which the gases and dust emanate to create the coma and the tails.

nucleus of an atom: The inner and dense part of an atom containing protons and neutrons.

Oort Cloud: The hypothesized region in the far outer reaches of the solar system that is required to explain the observations of the long-period comets.

opposition: The time in an outer planet's orbit when it is opposite the Sun on the sky.

Pangea: The name for the supercontinent, composed of all the continental plates, that existed on Earth about 200 million years ago.

parallax: The effect in which a nearby object appears to move against the reference frame of background stars as Earth orbits the Sun.

perihelion: The point in an orbit closest to the Sun.

photosphere: The visible surface of the Sun.

planet: This term has had various definitions over the years (see Lecture Nineteen). In 2006, the IAU defined a planet in our solar system as an object that (1) orbits the Sun, (2) is too small for nuclear fusion, (3) is large enough for gravity to make it spherical, and (4) dominates its orbit.

planetesimal: A rocky body in the forming solar system, larger than about 1 km, which is taking part in the accretion of the planets.

plate tectonics: The motion of the continental and oceanic plates across the surface of Earth and its consequences.

precession: The wobbling motion of a rotating object, such as a top, a planet, or a planet's orbit around the Sun. The rotation axis may change, or precess, in a cyclical manner over the course of many rotations.

prism: A device for bending light and splitting it into its component colors of the spectrum.

prominence: A long streamer of gas in the corona of Sun, often many hundreds of thousands of kilometers long. A prominence is a streamer that is seen extending off the edge of the Sun, whereas a **filament** is a streamer seen projected onto the face of the Sun.

proplyd: A contraction of "protoplanetary disk."

pulsar: A rapidly spinning neutron star that emits radio waves from its poles and is observed as short pulses of radio emission when the poles sweep through our line of sight.

quadrant: An astronomical observing tool that predates the telescope and is used for measuring the angle of a star above the horizon.

quasar: A contraction of "quasi-stellar radio source." An object that is star-like in appearance, can be observed by radio telescopes, and is located well beyond our galaxy.

radiative zone: The region of the Sun's interior where energy is transported by the emission and absorption of photons. See also **convective zone**.

radio wave light: The longest wavelength and lowest energy light of the electromagnetic spectrum.

reflecting telescope: A telescope that gathers light using mirrors.

refracting telescope: A telescope that gathers light using lenses.

regular moon: A moon of a planet that is believed to have formed along with the planet.

resolution: The minimum angular size that a telescope can see clearly. It is limited by the optics of the telescope and by the observing conditions, especially the atmosphere. Also called "resolving power."

resonance: In orbits, a repeating gravitational interaction between bodies because of the whole-number ratios of their orbits. For example, Pluto is in resonance with Neptune because it makes three orbits around the Sun in the same time it takes Neptune to make two orbits.

retrograde motion: The apparent backward motion, relative to the stars, seen for an outer planet near opposition.

right ascension: The measurement of angular distance on the sky in the east-west direction. Right ascension is measured with respect to a defined meridian. See also **declination**.

rocky planet: One of the four planets in the solar system made mostly of rock: Mercury, Venus, Earth, and Mars.

scarp: A discontinuity in the landscape created by geologic movement of the surface.

scattered disk objects: Kuiper Belt objects that have been shifted onto larger orbits by interaction with Neptune.

sedimentary rock: Rock that has been deposited by wind, water, or ice over time.

seismology: The study of earthquakes and the waves they generate.

semi-major axis: One half of the longer axis of an ellipse.

short-period comet: A comet that orbits the Sun in fewer than 200 years.

sidereal day: The average time between successive crossings of a star through the meridian: 23 hours, 56 minutes. Compare with **solar day**.

Small Magellanic Cloud: The smaller of two prominent dwarf galaxy companions of the Milky Way Galaxy. The SMC, as it is called, is about 200,000 light-years away and is only visible to observers in the southern hemisphere on Earth.

solar day: The average time between successive crossings of the Sun through the meridian: 24 hours. Compare with **sidereal day**.

solar flare: A sudden, intense release of magnetic energy from the Sun.

solar wind: A continuous stream of charged particles that flows away from the Sun.

solstice: The two points in Earth's orbit, during summer and winter, when the lengths of day and night show their greatest disparity.

spectral line: A thin bright or dark region in a spectrum that is created by the emission from a specific element or ion.

spectrum: The light from a star or other object dispersed into its constituent colors and wavelengths for study.

spicules: Thin, transient jets of hot gas in the chromosphere of the Sun. They are several thousand kilometers long and are constantly forming and dissipating on timescales of 5 to 10 minutes.

star: An object large enough to sustain hydrogen fusion at its core; an object with greater than roughly 70 times the mass of Jupiter.

stratosphere: The second layer in Earth's atmosphere, in which ozone molecules absorb ultraviolet light and the temperature increases with height.

subduction: In plate tectonics, the process of one plate going underneath another at a plate boundary.

sublimation: The direct transformation of solid to vapor, without going through a liquid phase.

sunspot: A dark region in the photosphere of the Sun.

super-Earth: An extrasolar planet with a mass several times that of Earth.

supernova: The explosion of a star.

termination shock: The point inside the heliopause at which the solar wind slows down from supersonic to subsonic speeds.

tesserae: Old regions on Venus's surface that show complex deformation.

theory: In science, a hypothesis that makes predictions, and whose predictions have been tested rigorously enough for near universal acceptance. In the public vernacular, this term has become essentially a synonym to "hypothesis."

thermosphere: The fourth layer in Earth's atmosphere, in which high energy radiation from the Sun is absorbed and the temperature increases with height.

tidal force: The stress created within a body when the gravitational pull on one side of the object is significantly stronger than the gravitational pull on the opposite side.

topography: A measurement or map of elevation.

transit: The passage of a small body in front of a larger body, such as a planet in front of a star. It is like an eclipse, but with the angular size of the background object much larger than that of the foreground object.

transition region: The thin (100-km) region of the Sun's atmosphere above the chromosphere and below the corona.

trans-Neptunian objects: A term roughly equivalent to Kuiper Belt objects.

Trojan Asteroids: Small objects, more comet-like than asteroid-like, that orbit the Sun both ahead of and behind Jupiter in its orbit.

troposphere: The lowest level of Earth's atmosphere, where energy balance is dominated by heat absorbed and reradiated from the ground.

Tunguska event: A massive explosion in the skies of Siberia on June 30, 1908.

ultraviolet light: The region of the spectrum just beyond violet light. Ultraviolet light waves have shorter wavelength and more energy than visible light waves.

visible light: The region of light that human eyes can see. Visible light has longer wavelengths than gamma ray, X-ray, and ultraviolet light, and shorter wavelengths than the infrared, microwave, and radio regions of the spectrum.

volatility: A measure of how much a substance will break down in the presence of heat. Ices are much more volatile than rocks.

X-ray light: A high energy and short wavelength region of light in between ultraviolet light and gamma rays.

zodiac: The traditional 12 constellations through which the Sun passes during the year. The concept of the zodiac is no longer accurate, given that the modern definition of constellation boundaries gives them unequal parts of the Sun's path. The Sun also passes through the constellation Ophiuchus, which is not part of the traditional zodiac.

Biographical Notes

Adams, John Couch (1819–1892). An English mathematician and astronomer, Adams was the first person to predict the existence of Neptune based on the irregularities in the motions of Uranus. Unfortunately, he failed to publish his result, and British astronomers did not search in earnest for the planet before a French astronomer also predicted the planet and a German observer discovered it. Adams is often credited as a codiscoverer nonetheless. Adams came from a family of little means, but his brilliance in mathematics enabled him to graduate at the top of his class at Cambridge. After gaining fame for Neptune at an early age, Adams continued doing mathematical research pertaining to astronomy and teaching at Cambridge for the rest of his life.

Aristarchus of Samos (c. 310–230 B.C.E.). Sometimes referred to as an ancient version of Copernicus, Aristarchus is the first in the historical record to propose a Sun-centered solar system. His views on the subject were dismissed by Archimedes, and no copies of his book survive. He is perhaps better remembered for his mathematical work on estimating the sizes of and distances to the Moon and the Sun, in a book that does survive. He is believed to have studied in Aristotle's Lyceum in Alexandria and to have done his work on the Moon and Sun early in his career, before he adopted the Sun-centered view. There is also some evidence that Aristarchus was mainly interested in the mathematical techniques for the calculations and not particularly in getting accurate observations or answers.

Aristotle (384–322 B.C.E.). It is difficult to overstate Aristotle's influence on astronomy, science, philosophy, and culture in Europe from ancient times through the Renaissance. The 30 works attributed to him span a range of subjects, and many of them were considered definitive. Certainly his view of the heavens as perfect and unchanging, in contrast to the base and fluctuating sphere of Earth, influenced and even governed astronomical thought for about 2000 years. He began studying at Plato's Academy in Athens at the age of 17 and proceeded to stay on as a teacher for 20 years. He left around the time of Plato's death and later founded his own school, the Lyceum. His school brought critical thinking to a broad range of studies, many of which had not been addressed previously. The lecture notes of Aristotle's that survive were not published during his lifetime but about 300 years later by the last head of the Lyceum.

Brahe, Tycho (1546–1601). The Danish astronomer Brahe is acknowledged as the greatest of the pre-telescopic observers. His observations, combined with Johannes Kepler's mathematics, were the lynchpin in providing astronomers with practical benefits of Copernican model: the improved predictive power of the *Rudolphine Tables*. Noble by birth, Brahe began his studies at the University of Copenhagen at just 13 years old. A year later, the successful prediction of an eclipse intrigued him, and he began studying astronomy. In his many travels as a young man, he learned both observing techniques and instrument making. Under the patronage of the king of Denmark, he built the finest observatory of the day and continually worked to improve his observing instruments. His observations of a supernova and a comet were able to show that objects beyond Earth definitely changed, a break with Aristotle's philosophy. He also created an alternative model of the solar system, in which the Sun and Moon orbited Earth but the planets orbited the Sun. After 20 years of royal support, he fell out with the new king of Denmark and took the position of imperial mathematician to the Holy Roman Emporer, Rudolph II, in Prague. Kepler was hired as an assistant to help work on the *Rudolphine Tables*, but Brahe died a premature death a year later.

Copernicus, Nicolaus (1473–1543). Although universally recognized as the astronomer who shifted our view of the solar system from Earth-centered to Sun-centered, the Polish-born Copernicus was officially employed as a canon at Frauenberg Cathedral. The relatively few duties of this appointment allowed one of the great scientific works in history to be pursued as a "hobby." In his youth, Copernicus began his study of astronomy while at the University of Krakow, and he continued it while in Italy, studying canon law in Bologna and medicine in Padua. Upon returning to his homeland, he served as physician to his uncle, the bishop, for five years before resuming his post as canon. In 1514, he distributed to a few friends a short work, the *Little Commentary*, that set down the principles of the Sun-centered universe. Over the subsequent years, warring states and administrative appointments left him little time for astronomy for nearly a decade. When he did find the time for research and writing a more detailed manuscript, it was another 20 years before it was published. His apparent reluctance to publish was overcome with the encouragement of a young astronomy professor, Georg Rheticus, who both publicized Copernicus's ideas and assisted in getting his book printed. It is said that Copernicus lived just long enough to see a copy of his masterpiece, *On the Revolutions of Heavenly Spheres*, on his deathbed.

Einstein, Albert (1879–1955). One of the most influential figures of the 20th century, Einstein helped herald a new era in modern physics with his special and general theories of relativity. These ideas provide the basis for interpretation and explanation of any number of astronomical observations, including the precession of the perihelion of Mercury and gravitational lensing. Einstein was educated in Munich, where contrary to a popular myth he was a fine student. He did, however, later fail an entrance exam for an engineering school and have trouble finding a university post after eventually getting his degree. He worked in a patent office for the better part of a decade, and while there he wrote three landmark papers in physics, one of which earned him a Nobel Prize. After returning to academia, Einstein embarked on a celebrated career at the forefront of physics and astronomy. He moved to the United States in 1932 and continued his research in Princeton.

Eratosthenes of Cyrene (276–194 B.C.E.). Eratosthenes is remembered in astronomy for his geometrical measurement of the size of Earth using shadows. His work on the subject was lost, but accounts by others mention it along with his work on the distances of the Sun and Moon, the tilt of Earth's axis, and the development of a calendar with leap years. He was a well-rounded scholar who did important work in other fields such as mathematics and geography. Born in North Africa, he also studied in Athens and succeeded one of his teachers as the third librarian at the famed library at Alexandria.

Galilei, Galileo (1564–1642). Galileo's status as the first telescopic astronomer is perhaps only exceeded by his reputation as the father of modern science. The latter reputation is based on his exacting critical thinking and demonstration that experiments can be used to test hypotheses. In astronomy, however, after his initial observational advances, his contributions shifted more to the philosophical in his determined arguments for the Copernican model. Educated by monks, Galileo studied medicine and mathematics at the University of Pisa. He gave up medicine in favor of math, eventually becoming a professor at Pisa and then at Padua. In Padua, his interest in astronomy became apparent, leading to his improvement of spyglass optics to create a suitable telescope in 1609. His discoveries with the telescope enabled him to obtain a post with the Grand Duke in Florence. His 1616 *Letter to the Grand Duchess* affirmed his belief in the physical reality of the Sun-centered universe but led to the Catholic Church's forbidding him to espouse Copernican views. Though he had several audiences with the Pope over the next decade, his vigorous arguments in favor of the banned Copernican system in his 1632 book, *Dialogue Concerning the Two Chief World Systems*, were not overlooked by the Inquisition. He was forced to admit he was mistaken and was sentenced to house arrest for the rest of his life.

Herschel, William (1738–1822). Brought up as a musician in Germany, Herschel moved to London at the age of 19. He continued teaching, performing, and conducting music while also developing his interests in mathematics and astronomy. He was especially adept at making telescopes, constructing over 400 in his lifetime. The quality of his instruments enabled him, in 1781, to identify a comet-like object while doing a survey of double stars. Upon deeper investigation by professional astronomers, the planet Uranus was recognized, and Herschel achieved notoriety that enabled him to pursue astronomy full-time. His later work in conjunction with his sister Caroline discovered moons, catalogued nebulae, and investigated the motions of stars. He also discovered infrared radiation, coined the term "asteroid," and helped found what would become the Royal Astronomical Society. His only son, John, became an important astronomer as well.

Kepler, Johannes (1571–1630). Kepler's intensive search for order led to both his greatest admiration and his strongest criticism. A devoutly religious man, he felt that the universe must be structured according to mathematics, and thus much of his work is filled with what we would now call "numerology." However, this insistence on looking for number patterns everywhere allowed him to uncover three empirical laws that were the first correct description of planet orbits. Born in what is now Germany, he was educated at the University of Tubingen, where he went beyond the traditional astronomy and became convinced of the physical truth of the Sun-centered system. His 1596 solar system model explained the six planets' orbits of the Copernican model using spheres circumscribing the five convex regular solids. Hired to assist Tycho Brahe in 1600, he took over as imperial mathematician to the Holy Roman Emperor, Rudolph II, when Brahe died a year later. Access to Brahe's accurate observations led him into several years of "war with Mars" that resulted in two of his three laws in the book *Astronomia nova* (1609). He also published on the supernova of 1604, studied the properties of lenses, and made a major improvement on Galileo's design of the telescope. His third law came after moving to Linz and late in the production of his intensely numerical *Harmonies of the World* (1619). His most important practical work, the *Rudolphine Tables* (1628), was published nearly 30 years after he was hired to assist Brahe in its creation.

Le Verrier, Urbain (1811–1877). A gifted French mathematical astronomer, Le Verrier was successful in both predicting the existence of Neptune and in getting someone to search for it. In this, he rose above his unknown English rival, John Couch Adams, to receive primary credit for its discovery. His education, first in Normandy and eventually at the École Polytechnique in Paris, was marked by a strong intellect and diligent work. Le Verrier started his career in chemistry but abruptly switched to astronomy when a position opened up on the faculty of the École Polytechnique in 1837. His facility with mathematics led him to much success in celestial mechanics, the calculation of planetary orbits. The prediction of Neptune gave him honors and prestige, and he eventually became director of the Paris Observatory. Such was his reputation that his prediction of a planet or planets inside Mercury became immediately probable and prompted many searches. Only four decades after his death would Albert Einstein's general theory of relativity correctly explain the discrepancy in Mercury's orbit.

Newton, Isaac (1643–1727). Newton, an Englishman, stands as one of the most important figures in mathematics and physics for his inventions of mathematical techniques (including the calculus) and his application of them to explain physical phenomena. His universal law of gravity provides the fundamental mathematical basis that transformed solar system calculations from descriptive to predictive. Newton's father died three months before he was born, and his youth was unsettled and not marked by academic distinction. Only when he was preparing for and studying at Cambridge did his intellectual abilities show themselves, including his special depth in mathematics. Sent home during a plague, he proceeded to develop new ideas in math, physics, and astronomy that led to a professorship in 1669 at the remarkable age of 27. His most productive period then followed, with studies in optics, the laws of motion, and his magnum opus, the *Principia*, in 1687. Newton's fame, however, drew him into more political arenas, and his scientific work became less productive. In 1693, Newton gave up research altogether and moved into government positions, including master of the mint. He remained active in, and became president of, the Royal Society (the preeminent science organization), published some of his earlier research, and was the first scientist knighted by the Queen.

Ptolemy, Claudius (c. 85–165). The Greek astronomer and geographer Ptolemy is arguably the most influential astronomer of all time, having had his work remain the standard text for some 1500 years. His compendium of astronomical knowledge organized, improved, and presented the Earth-centered model of the solar system that had been developed over more than 500 years of study. Relatively little is known about his life, except that his scientific work was done in Alexandria and that he published treatises on geography, optics, music, and astrology in addition to his astronomical standard, the *Almagest*. Much discussion about Ptolemy thus diverges into scientific intrigue as to whether he developed much new work himself, properly credited those whose work he encapsulated, or even fabricated some of his star catalog data. The most important evidence, his works, indicates that he was learned in several fields and, at the very least, a masterful expositor of science.

Bibliography

Readings:

Bally, John, and Reipurth, Bo. *The Birth of Stars and Planets*. Cambridge: Cambridge University Press, 2006. A nontechnical discussion of all aspects of star and planet formation.

Beatty, J. Kelly, Carolyn Collins Petersen, and Andrew Chaikin. *The New Solar System*. 4th ed. Cambridge, MA: Sky Publishing, 1999. Excellent collection of chapters by experts. The best companion to this course, although an updated fifth edition would be welcomed.

Bell, Jim. *Postcards from Mars*. London: Dutton, 2006. A collection of amazing images from the Mars rovers, with discussion by a lead scientist on the mission.

Bennett, Jeffrey, Megan Donahue, Nicholas Schneider, and Mark Voit. *The Cosmic Perspective*. 5th ed. New York: Addison-Wesley, 2008. A solid undergraduate text.

Benson, Michael. *Beyond: Visions of Interplanetary Probes*. New York: Abrams, 2008. An exquisite picture book of satellite and space probe images.

Berlinski, David. *Newton's Gift: How Sir Isaac Newton Unlocked the System of the World*. New York: Touchstone, 2000. A history of Newton mixed with the mathematical development of his laws of motion.

Burham, Robert. *Great Comets*. Cambridge: Cambridge University Press, 2000. A pretty full description of observational, historical, and some scientific ideas about this too-often ignored class of objects.

Casoli, Fabienne, and Therese Encrenaz. *The New Worlds: Extrasolar Planets*. Chichester, UK: Praxis Publishing, 2007. A compact and readable book on the first decade of discovery in this new field.

Comins, Neil F. *What If the Moon Didn't Exist? Voyages to Earths That Might Have Been*. New York: HarperCollins Publishers, 1993. Some intriguing thoughts on the Moon's influence on Earth and other scenarios. May be out of print.

Copernicus, Nicolaus. *On the Revolutions of Heavenly Spheres*. Translated by Charles Glenn Wallis. Amherst, NY: Prometheus Books, 1995. The hypothesis of the Sun-centered solar system. The philosophical arguments of Book One are the best part. The geometrical and mathematical arguments are dense.

Croswell, Ken. *Planet Quest: The Epic Discovery of Alien Solar Systems*. New York: The Free Press, 1997. The first half-decade when planets began to be discovered around other stars, told with sufficient scientific detail to allow appreciation of the magnitude of the efforts.

Crowe, Michael J. *Theories of the World from Antiquity to the Copernican Revolution*. Mineola, NY: Dover Publications, 2001. Excellent examination and discussion of pre-telescopic astronomy from the Greeks to Galileo.

Davies, John. *Beyond Pluto*. Cambridge: Cambridge University Press, 2001. A book that covers the emergent knowledge of the outer solar system and provides scientific perspective from before the Pluto debate became public.

De Pater, Imke, and Jack J. Lissauer. *Planetary Sciences*. Cambridge: Cambridge University Press, 2001. A graduate-level text in solar system studies. Very technical and mathematical, as the intended audience consists of those aspiring to be professional astronomers.

Dessler, Andrew E., and Edward A. Parson. *The Science and Politics of Global Climate Change*. Cambridge: Cambridge University Press, 2006. A reasoned and accessible look at not only the observed problem but also the practical obstacles to a solution.

Ferguson, Kitty. *Tycho & Kepler*. New York: Walker Publishing, 2002. An account of the two astronomers' lives and their combined scientific legacy.

Feynman, Richard P. *The Pleasure of Finding Things Out*. New York: Basic Books, 1999. A collection of short works by a brilliant physicist. The title essay captures the essential heart of scientific inquiry.

Galilei, Galileo. *Dialogue Concerning the Two Chief World Systems*. Translated by Stillman Drake. New York: The Modern Library, 2001. The work for which Galileo appeared before the Inquisition.

———. *Siderius Nuncius*. Translated by Albert Van Helden. Chicago: University of Chicago Press, 1989. Galileo's very readable initial discoveries with the telescope.

Gingerich, Owen. *The Book Nobody Read*. New York: Walker & Company, 2004. An interesting sideways look at Copernicus and his times through examination of the surviving copies of his book.

Golub, Leon, and Jay M. Pasachoff. *Nearest Star: The Surprising Science of Our Sun*. Cambridge, MA: Harvard University Press, 2002. A look at all aspects of our Sun both as a star and as part of the solar system.

Grosser, Morton. *The Discovery of Neptune*. New York: Dover Publications, 1979. Out of print. The detailed and fascinating story of the astronomy, personalities, politics, and intrigue behind the planet that was predicted by mathematics before being discovered with a telescope.

Hakim, Joy. *The Story of Science*. Washington, DC: Smithsonian Books, 2004. A historical overview of the development of scientific thought aimed at younger readers.

Hey, Nigel. *Solar System*. London: Weidenfeld & Nicolson, 2005. Somewhat uneven but has some good essays on current topics.

Hill, Steele, and Michael Carlowicz. *The Sun*. New York: Abrams, 2006. A book of incredible pictures from the SOHO mission, supplemented with several short essays.

Hoyt, William Graves. *Planets X and Pluto*. Tucson, AZ: University of Arizona Press, 1980. A detailed look at the many predictions and searches for planets in our solar system.

Impey, Chris. *The Living Cosmos*. New York: Random House, 2007. A detailed look at the search for life and astrobiology.

Jewitt, D., A. Morbidelli, and H. Rauer. *Trans-Neptunian Objects and Comets: Saas-Fee Advanced Course 35*. Berlin: Springer, 2007. A graduate-level text on the outer solar system from three experts in the field.

Levy, David H. *Clyde Tombaugh: Discoverer of Planet Pluto*. Cambridge, MA: Sky Publishing, 2006. A history of Tombaugh's life, with good chapters on the Pluto search.

Lewis, John S. *Rain of Iron and Ice*. New York: Addison-Wesley Publishing, 1996. Excellent discussion of asteroids, impacts, and the threat they pose.

Light, Michael. *Full Moon*. New York: Alfred A. Knopf, 1999. An exquisite collection of Apollo moon photographs.

Lovett, Laura, Joan Horvath, and Jeff Cuzzi. *Saturn: A New View*. New York: Harry N. Abrams, 2006. Features amazing images of Saturn from the Cassini mission.

Mackenzie, Dana. *The Big Splat, or How Our Moon Came to Be*. Hoboken, NJ: John Wiley and Sons, 2003. A history of Moon observations and ideas leading to the current giant impact formation hypothesis.

McFadden, Lucy-Ann, Paul Weissman, and Torrence Johnson. *Encyclopedia of the Solar System*. 2nd ed. San Diego, CA: Elsevier, 2007. A reference book of essays by experts on most of the topics covered in this course.

Moore, Patrick, and H. J. P. Arnold. *Space: The First 50 Years*. New York: Sterling, 2007. Both a picture book and a well-written selective history of the space age.

O'Dell, C. Robert. *The Orion Nebula: Where Stars are Born*. Cambridge, MA: Harvard University Press, 2003. An expert describes all aspects of his favorite object of research.

Ptolemy, Claudius. *Almagest*. Translated by G. J. Toomer. Princeton, NJ: Princeton University Press, 1998. The introductory materials provide authoritative historical perspective, and the dense mathematical arguments are interesting to peruse.

Sagan, Carl. *The Demon-Haunted World: Science as a Candle in the Dark*. New York: Ballantine Books, 1997. An exploration of scientific thought and the scientific method as applied to diverse subjects.

Shu, Frank H. *The Physical Universe: An Introduction to Astronomy*. Mill Valley, CA: University Science Books, 1982. A college textbook for science majors and those comfortable with mathematics. A bit out of date and without color pictures, but the exactness of the arguments is generally not matched in other college texts.

Sobel, Dava. *The Planets*. New York: Penguin, 2006. A poetic but scientifically accurate look at the solar system.

Sullivan, Woodruff T., III, and John A. Baross. *Planets and Life*. New York: Cambridge University Press, 2007. An excellent collection of research-level discussion on nearly all topics in astrobiology. Some chapters are readable, while many parts are highly technical and dense with chemistry.

Verschuur, Gerrit L. *Impact! The Threat of Comets and Asteroids.* New York: Oxford University Press, 1996. A good discussion of the history of impacts and the way scientists look for and consider potential impact threats.

Villard, Ray, and Lynette Cook. *Infinite Worlds: An Illustrated Voyage to Planets Beyond Our Sun.* Berkeley, CA: University of California Press, 2005. The combination of a knowledgeable writer and imaginative art work in a sweeping view of how planetary systems develop.

Voelkel, James R. *The Composition of Kepler's "Astronomia nova."* Princeton, NJ: Princeton University Press, 2001. A heavily detailed analysis of Kepler's most important work.

Weintraub, David A. *Is Pluto a Planet?* Princeton, NJ: Princeton University Press, 2007. An excellent history of planetary and solar system discoveries.

Wilson, Robert. *Astronomy through the Ages.* Princeton, NJ: Princeton University Press, 1997. A good history with more science than is generally found in such texts, but also a little dry.

Internet Resources:

Ancient Planetary Model Animations. Dennis Duke's interactive Flash animation models of various Earth-centered and Sun-centered models of the solar system. http://people.scs.fsu.edu/~dduke/models

"Astronomy Picture of the Day." A timely and convenient way to explore astronomy through beautiful and interesting images. http://apod.nasa.gov/

Cassini-Huygens. All of the details of this groundbreaking mission to Saturn. http://saturn.jpl.nasa.gov/

Definition of Planet at the 2006 IAU General Assembly. International Astronomical Union. Resolutions 5 and 6 on this page show the 2006 IAU definition of planet. http://www.iau.org/Resolutions_at_GA-XXVI.340.0.html

Dwarf Planets. By Mike Brown. Discussion from one of the main discoverers of Kuiper Belt objects. http://www.gps.caltech.edu/~mbrown/dwarfplanets.html

Eclipse Web Site. NASA. More information about eclipses than you ever wanted to know. http://eclipse.gsfc.nasa.gov/eclipse.html

Exoplanets. From the California & Carnegie Planet Search. Information of planets around other stars from a leading research group. http://exoplanets.org/

The Extrasolar Planets Encyclopaedia. An updated catalog of information and data about planets around other stars. http://exoplanet.eu/

Frank Summers. Astronomy, visualizations, and other information by and about the course professor. http://terpsichore.stsci.edu/~summers/

Galileo Journey to Jupiter. The archived results of this incredibly successful mission to Jupiter. http://www2.jpl.nasa.gov/galileo/

The Galileo Project. Rice University. An excellent resource on Galileo's life and works. http://galileo.rice.edu/

HubbleSite. The website for news, images, and background on the Hubble Space Telescope. http://hubblesite.org/

Kuiper Belt. By Dave Jewitt. Discussion, updates, and opinion from one of the leaders in the field. http://www.ifa.hawaii.edu/faculty/jewitt/kb.html

MacTutor History of Mathematics Archive. Excellent biographies of mathematicians and scientists, plus considerable historical material including mathematical astronomy. http://www-history.mcs.st-andrews.ac.uk/

Mars Exploration Program. NASA's website for keeping up with the many and continuing missions to the red planet. http://mars.jpl.nasa.gov/

Messenger: MErcury Surface, Space ENvironment, GEochemistry, and Ranging. The first mission to explore Mercury in decades will begin orbiting the planet in 2011. http://messenger.jhuapl.edu/

NASA History Division. Details of NASA missions and operations including some exceptional archival material. http://history.nasa.gov/

New Horizons. The first mission to explore Pluto and the Kuiper Belt will begin in earnest in 2015. http://pluto.jhuapl.edu/

The ~~Nine~~ 8 Planets. A well-assembled collection of information about the solar system. http://www.nineplanets.org/

Photojournal. The most complete collection of NASA solar system images available, generally with the accompanying press release text. http://photojournal.jpl.nasa.gov/

SOHO Solar and Heliospheric Observatory. Excellent scientific visuals of the Sun from the Solar and Heliospheric Observatory. http://sohowww.nascom.nasa.gov/

Visible Earth. Satellite images of our home planet from NASA. http://visibleearth.nasa.gov/

Notes